Virtual Reality Designs

Editors

Adriana Peña Pérez Negrón
Universidad de Guadalajara
Guadalajara, México

Graciela Lara López
Universidad de Guadalajara
Guadalajara, México

Héctor Rafael Orozco Aguirre
Universidad Autónoma del Estado de México
Estado de México, México

CRC Press
Taylor & Francis Group
Boca Raton London New York

CRC Press is an imprint of the
Taylor & Francis Group, an **informa** business
A SCIENCE PUBLISHERS BOOK

CRC Press
Taylor & Francis Group
6000 Broken Sound Parkway NW, Suite 300
Boca Raton, FL 33487-2742

© 2020 by Taylor & Francis Group, LLC
CRC Press is an imprint of Taylor & Francis Group, an Informa business

No claim to original U.S. Government works

Version Date: 20200210

International Standard Book Number-13: 978-0-367-89497-9 (Hardback)

Visit the Taylor & Francis Web site at
http://www.taylorandfrancis.com

and the CRC Press Web site at
http://www.crcpress.com

Printed and bound by CPI Group (UK) Ltd, Croydon, CR0 4YY

Preface

Virtual Reality Designs

Presentation

Due to technological advances, we are currently living realities which were never ever imagined; particularly through Virtual Reality (VR), it is possible to visit distant places without moving and even interact with virtual people. As Steuer (1995) pointed out, VR is an alternate computer-generated world that responds to human interaction. Virtual Reality is the technology to create a whole digital environment, and it is increasingly entering the life of regular computer users.

Any experience, as simple as it might be, could be outstanding if generated under the VR technique. But also, activities that might not be possible to carry out in real life can now be performed virtually, like, for example, flying or jumping an abyss without consequences. A key objective for VR is to provide a sense of presence (Azmandian et al., 2016), but also, this sense could be by personifying an insect or a giant. VR users can explore or develop activities that situate them on a new way of communication with others, working, playing, and perhaps, falling in love.

Another characteristic of VR is immersion, where the user can interact solely with the virtual world; in this case, the user can be transported to the digital scenario through glasses and headphones to be isolated from the real world in a 360° view. Computer improvements in graphic management and the drop in the cost of VR devices have permitted many users to get immersive VR at home. Undoubtedly, in recent years, VR has become one of the greatest technological trends, representing a great revolution in the technological industry.

Over the years, users have experienced VR in various sectors; here are some examples of the fields that VR has impacted:

- *Education*: VR has been used to facilitate teaching, usually with positive results, by changing the education methodology by combining technological tools and educational methods. In education, VR entails motivational aspects to capture the attention of the students, so as to contribute to better results.

- *Travel and Hospitality Agencies*: With the help of VR, it is possible to provide users with the experience of interacting with a set of services offered by lodging and food companies to travelers.

- *Real Estate*: Undoubtedly one of the sectors where VR has been strategically useful is the real estate. Here VR offers the possibility of saving time and money by allowing appreciation of the details of the houses, such as size, textures, colors, and materials before physically changing them. It turns out to be quite feasible in recreating the real estate in a three-dimensional (3D) animation.

- *Automotive*: The driving or automotive experience via VR has favored buyers of the large automobile manufacturers by offering the possibility of driving one of their promoted models before purchase.
- *Journalism*: *The New York Times* is one of the newspapers that is currently producing immersive journalism, presenting news as if readers were living it from the scene (through the support of app nytvr).
- *Electronic Commerce*: With the support of VR, it is possible to tour an online store. This has proved to be very attractive to consumers who enjoy doing so from home. Also, with a well-designed virtual environment, it is possible to customize products, combine colors, and to know products' availability in advance. Likewise, the so-called 'businesses on the street' is a service that is proving quite innovative, particularly for small businesses as it showcases the business and their merchandise through VR. This brings buyers from anywhere in the world to virtually visit the establishments.
- *Health*: VR has been used for the management of various pathologies; among the most outstanding, is the management of phobias and anxiety. In surgical operations, it is already a reality to virtualize parts of the human body to be examined in depth. VR has also been the means for training future medicos.
- *Leisure*: Nowadays one of the most popular technological trends in VR is videogames. Through stories, challenges, and adventures, users are submerged in these electronic games only for the pleasure of experiencing sensations that provoke emotions.
- *Concert and Festivals*: Nowadays it is possible to offer concerts or mass festivals in which the number of attendees is surprisingly large; virtual attendees experience the event as real, as if they were assembled on it.

When a user interacts with VR to watch a 3D environment with expected ease of use, to freely move or navigate and to take care of activities that can be practiced individually or collectively. This user experience should be supported by a virtual interface developed under a human-computer-interaction (HCI) approach (Estrada, 2008). In order to enhance the user experience, for the VR as for any other software, a key feature is the environment design. The design is an essential stage, prior to and during the development of an application, although it does not always receive proper attention.

Samperio et al. (2018) presented a methodology for modeling VR systems for learning, on mobile devices, a methodology that might well fit into the development of any VR application. According to Samperio et al. (2018), the design of a virtual environment is the stage where different activities are considered—from the description of scenarios, the analysis of the needs and the definition of the goals to be met. Subsequently, the implementation of the mechanisms, that allow achievement of the application objectives, is carried out. Likewise, the specification of the functions to be executed in the virtual environment is performed followed by the realization of technical specifications.

A VR-environment design requires consideration of several factors—the first of which has to be the kind of activity that the user will carry out on it. Starting from there, it is also important to take into account the users' perception that will necessarily generate in him/her feelings and reactions to the environment.

Humans perceive ideas or things based on the information that their brain relates through their senses combined with previous knowledge, experiences, tastes, and personal instincts. The result of this process is actually what the person observes. That is why perception becomes a primary element to distinguish reality from virtuality (Díaz, 2007).

In this book are presented a range of studies linked to the user behavior, and the VR design that will guide to different user experiences in VR. Likewise, the applications developed for its research and analysis are related to activities of our daily life, in which different topics are addressed.

Chapter 1: Virtual Reality, as a Tool for Behavior Analysis and Applications by Rangel, Peña and Torres, brings to our attention the benefits of experimental control that VR offers for human behavior studies.

Human behavioral sciences, such as psychology and neuroscience, have renewed their experimentation techniques to better understand the reactions of human behavior by considering VR that simulates contexts in a controlled situation and offers the possibility of interaction among users.

In simulated digital situations, researchers can watch people's behavior, preventing them from experiencing the sensation of being observed. Through VR, the users can be surrounded by their work environment, a festive or an unexpected situation, and diverse contexts to understand how they confront such situations (Jofré et al., 2016).

In the experiences perceived within VR technology, the configuration of the environment would influence the users' behavior, particularly their capacity for altering the scenario—the central idea discussed in this chapter.

Chapter 2: Navigation in Virtual Reality by Peña, Lara and Estrada, addresses a fundamental function within VR applications. From the conceptual point of view, 'navigate' refers to the action of moving or getting around, through water, through a page or other documents in a computer network; that is, moving from one part to another.

VR navigation can be achieved in ways that in real life might not be possible at a time, such as teletransportation. Here, the users' types of movements to navigate a virtual environment can be by physical movement, manual viewpoint, steering, target-based travel, and route planning, among others, a navigation approach based on the design related to the events that will take place in the environment.

In this context, the authors relate elements of navigation in VR. They also describe the particularity of navigational aids in VR and the use of VR for spatial knowledge and training.

Chapter 3: User Modeling Systems Adapted to Virtual Environments by Lara, Peña, Paladines and Rubio, highlights the users' important role in the specifications of any VR system, since they are the ones to directly interact with the scenarios. However, not all users have the same background, preferences, knowledge or skills. The set of users' characteristics can be turned into rules for the VR environment to be adaptive for them.

The process of adjusting the environment to the users' characteristics is technically known as 'user modeling'—a characterization process that assumes that different types of users exist and they can be understood through taxonomy.

From a practical perspective, user modeling determines the design from the environment to adapt to the interaction of each user in the system, so that their participation can be totally focused on them and in the tasks or roles of the activity at hand. The users' characteristics on a set of established rules assist the development designer to understand, represent and validate the VR systems. This is also a helpful practice for users to enhance the VR for their particular use.

Another important element of VR is that artificial inhabitants can be generated to interact with humans or among them to create the digital world. The next three chapters deal with this challenging design issue.

Chapter 4: Synthetic Perception and Decision-making for Autonomous Virtual Humans in Virtual Reality Applications by Orozco, Thalmann and Ramos touches the interesting theme of creating virtual characters—an approach that is gaining interest in the study of humans beings.

People perceive the environment by observing movement, perspectives, others, and objects among a number of other elements to create a representation in a 3D context to make decisions and the selection of action among other possible ones. This chapter deals with the aspect of creating embodied virtual agents that simulate credible behavior when making decisions by a synthetic perception of the environment.

Authors also consider emotions as an influence on perception and decision-making processes, since people are basically emotional beings. Emotions move us to take certain decisions. This line of research that has aroused special interest is discussed to design virtual humans in a quasi-real context.

Chapter 5: An Internal Model for Characters in Virtual Environments: Emotions, Moods, and Personality from Julca, Médez and Hervás focuses on the exciting area of creating emotion, mood, and personality for artificial humans. Detecting the emotions and personality of people was until recently a unique quality of humans.

VR is able to generate emotional states that can be useful in different psychological therapies, to achieve the improvement of certain emotional changes in people, and even understand their current state of mind. Likewise, the development of different models of emotional recognition in environments of VR can predict the emotions that can generate changes in people in a real context.

Through diffused logic, the authors designed a specialized system to calibrate emotions and moods according to personality traits linked to the virtual human goals.

Chapter 6: Pedagogical Agents as Virtual Tutors: Applications and Future Trends in Intelligent Tutoring Systems and Virtual Learning Environments by Orozco, explores the use of virtual agents aimed at assisting learning and offering guidelines for its design and describing the insights of the Mexican intelligent pedagogical agent for schoolchildren.

This is one of the most extensive research areas within the virtual agents' fields; however, there are still several open issues among which, one of them is the affective element, particularly on adapting it to the apprentice. But let us not forget about the many factors, such as the content, the number of students, proper feedback, among others that affect the learning outcome.

The use of VR in the study of human behavior has also opened the possibility to use it as a support to understand, not only personal behavior, but also team behavior. The next chapter proposes to explore this feature.

Chapter 7: Gamification in Virtual Reality Environments for the Integration of Highly Effective Teams by Muñoz, Peña and Hernández uses VR to integrate teamwork. The development of VR along with the theory and dynamics of games, also known as gamification, is increasingly integrated into companies within their business processes, marketing or human resources.

Gamification can be described as the use of game elements in a non-ludic context, which supports commitment generation, making the employees to produce better performance and promoting a better behavior.

In recent years, the use of gamification in the software industry has been potentially used for the formation of work teams. In this context, VR implementation with gamification is a powerful tool and an emerging technology that transports motivation and commitment into a series of functional results.

Given that the software industry depends on the good performance of human capital and that it is not always easy to combine different types of people to get a productive workflow, the authors propose the use of VR to immerse possible teammates in a game situation to diagnose their probable role in a software development team, thereby determining their best roles based on their team skills.

Creating virtual worlds is also important to study or generate crowds' behavior, which is the next chapter's topic.

Chapter 8: Integrating Virtual Reality into Learning Object-based Courses by Muñoz-Arteaga and Cardona presents the possibility of integrating VR into learning objects. A learning object is a reusable piece of broken parts of a subject content to fulfill an instructional objective, particularly for virtual education or e-learning.

For a learning object, VR brings real-life or hypothetical scenarios with advantages such as any place labs with unlimited materials. Authors present design issues to incorporate VR in learning objects using structural and architectural models. They also present a case of study in four courses.

Chapter 9: Virtual Simulation of Road Traffic Based on Multi-agent Systems by Orozco, Quintana, Lazcano and Landassuri generates the simulation of road traffic for decision-making. The control of the terrestrial traffic in big cities is challenging; in order to find a solution that prevents accidents or other types of problems generated by vehicular congestion, a virtual simulation is proposed as a helpful tool.

The Virtual Reality multi-agent system approach was tested for land traffic management as an efficient tool. A number of influencing factors were modeled, such as a vehicle, traffic lights, pick hours, parking lots, bus stops, and roads, through agents to simulate their behavior in order to understand what creates vehicular traffic.

By means of 3D simulations of sites, spaces, or urban areas with dense traffic, and virtual agents, the circulation of motorists can be improved avoiding chaos, increasing safety, reducing accidents, and waiting times for drivers and pedestrians.

Several ideas emerged on this subject with the particular objective of getting self-management of traffic. This allowed creating a fully computerized vision understanding the generated information to support decision-making in this regard.

Chapter 10: The Sense of Touch as the Last Frontier in Virtual Reality Technology by Martínez, García, Oliver, González and Molina deals with what they call the last frontier. VR is a mostly visual technology that incorporates audio, managing mainly two senses; a third sense that has been somehow undermined due to technical difficulties is touch.

One of the most desired experiences of users of the VR is to try to grab a virtual object. This experience remains as one of the greatest challenges of VR. Where undoubtedly, gloves are a breakthrough allowing users to feel the weight or texture of what they touch.

For researchers, these technological advances would generate new studies regarding emotions, and leading to new experiments.

Beyond the visual and auditory channels, in which there are already great technological advances, people expect that the Virtual Reality stimulates the touch, and then to make us feel the heat, cold or even pain; hopefully not in along-distance future.

Acknowledgement

We would like to acknowledge the peer-reviewer effort by Mirna Ariadna Muñoz Mata, Nora Edith Rangel Bernal, Carlos de Jesús Torres Ceja, Gonzalo Méndez, Graciela Lara López, José Pascual Medina Massó, Adriana Peña Pérez Negrón, Héctor Rafael Orozco Aguirre, Alfredo Zapata González and José Eder Guzmán.

References

Azmandian, M., Hancock, M., Benko, H., Ofek, E. and Wilson, A.D. (2016, May). Haptic retargeting: Dynamic repurposing of passive haptics for enhanced virtual reality experiences. *In*: Proceedings of the 2016 CHI Conference on Human Factors in Computing Systems, ACM.

Díaz Pier, M. (2007). Realidad Virtual Basada en Percepción, unpublished Ph. D. thesis, Tecnológico de Monterrey, México.

Estrada, F. (2008). Diseño gráfico y entornos virtuales. Revista Digital Universitaria, Revista Digital Universitaria, UNAM, México, 9(9).

Jofré Pasinetti, N., Rodríguez, G., Alvarado, Y., Fernández, J. and Guerrero, R.A. (2016, May). La realidad virtual en los comportamientos sociales. *In*: XVIII Workshop de Investigadores en Ciencias de la Computación (WICC 2016, Entre Ríos, Argentina).

Samperio, G.A.T., Arcega, A.F., Sánchez, M.D.J.G. and Navarrete, A.S. (2018). Metodología para el modelado de sistemas de realidad virtual para el aprendizaje en dispositivos móviles. Pistas Educativas, 39(127): 518–534.

Steuer, Jonathan (1995). Defining virtual reality: Dimensions determining telepresence. pp. 33–56. *In*: Bioca and Levy (eds.). Communication in the Age of Virtual Reality, Hillsdale: Lawrence Erlbaum.

Contents

Chapter **1**

Virtual Reality as a Tool for Behavior Analysis and Applications

Nora Edith Rangel Bernal, Adriana Peña Pérez Negrón
and *Carlos de Jesús Torres Ceja**

1. Introduction

As in other disciplinary fields, for psychological treatment and psychological research, computer tools and simulations have been increasingly incorporated. This fact has brought multiple advantages to researchers, professionals, and users—advantages, such as the possibility of presenting more attractive tasks and interfaces, generating greater interest in the users, and permanence in situations for study and/or intervention. Other advantages of the use of these tools in behavior analysis are related to a methodological control, and the facilitation of data collection and its subsequent analysis. Undoubtedly, in this field the use of non-immersive virtual reality systems is privileged, but given the aforementioned advantages, it is not difficult, nor risky, to say that the use of systems that imply higher degrees of immersion could increase the ecological value of the research in psychology without sacrificing the control that the studies require. They would even allow an approach to research contexts, that would otherwise have elevated costs, or in some cases, be practically impossible. Also, psychologically speaking, it could promote different and more complex interactions between individuals and their environment.

As in other fields of knowledge, professionals and researchers in the psychological and behavioral sciences are recognizing the potential and promissory use of technology as an effective tool for their professional practice. Technologies have proved its efficiency in clinical psychology and in behavioral basic research (Loomis et al., 1999). In this chapter, it will be described how technology, particularly Virtual Reality (VR), has been implemented in these psychological areas, the obtained results, as well as the advantages and limitations that such implementation has brought to the discipline. Also, an analysis about the impact and implications of using VR tools in the understanding of psychological behavior is discussed.

Universidad de Guadalajara, Mexico.
* Corresponding author: dejesus.torres@academicos.udg.mx

2. VR Uses in Behavioral Analysis

Virtual reality technology is a three-dimensional (3D) computer-generated display that allows or compels the users to get the feeling of 'being there' or being present in an environment other than the one they actually are in and to interact with that environment (Ellis, 1995).

In clinical psychology, applications using VR are supporting therapeutic interventions for the treatment of anxieties and phobias (Gutiérrez, 2002; García-García and Rosa-Alcazar, 2011; Pelissolo et al., 2012; Dos Santos et al., 2018), as well as for the treatment of other clinical disorders, like autism (Goldsmith and Le Blanc, 2004; Ploog et al., 2013; Stichter et al., 2014), stuttering (Brundage and Hancock, 2015), deficient social skills (McKenney et al., 2004), lack of adherence to treatments that promote healthy habits, like physical exercise (Calogiuri et al., 2018), obesity and eating disorders (Riva, 2011; Cesa et al., 2013), among others. In the practice, VR used as a tool of intervention showed similar effectiveness, sometimes even higher, when compared to *in vivo* treatments (Pelissolo et al., 2012; Ploog et al., 2013).

In specialized literature, there are several papers showing detailed reviews of such studies and interventions, as well as the description of their results and implications in the psychological and clinical practice (Powers and Emmelkamp, 2008; Pelissolo et al., 2012; Ploog et al., 2013).

According to Pérez Martínez (2011), VR helps to improve people's life quality in various fields of their daily activity, including science. In the case of behavioral sciences, the use of this technology has been varied—it has been used for data collection, self-monitoring, and immediate feedback for the users. However, its use has been limited in most cases, to the presentation of tasks on a computer screen with the main goal of getting the automatic record of the participant's responses, usually given through a mouse, a keyboard, or game control.

Some researchers have incorporated VR in their laboratory work, implementing the use of diverse gadgets like glasses, helmets, gloves or even suits that allow the subjects to interact directly with their environment, including other users (e.g., collaborative virtual environments). Thus, virtual reality can include visual, auditory, tactile and olfactory stimulation with the purpose of creating in the user a sense of reality (Brundage and Hancock, 2015).

Pérez Martínez (2011) suggested that the incorporation of VR in the scientific work can overcome the limitations of traditional research methods and non-immersive technology that hinder participant's movement in the study situation, limiting in this way, the type of interaction or the relation between the participants and their environment. 3D virtual environments offer to behavior researchers the possibility of developing more complex and enriched interactive settings, increasing the number and the quality of contacts that an individual can establish with his environment. This allows the possibility of studying active individuals–individuals in the movement that used to be spatially immobile people in the study situation (García-García and Rosa-Alcazar, 2011) to the extent that the individual is no longer passive. A dimension of movement, in real time, is incorporated into the analyzed situations, rendering individual's interaction more varied and complex.

A very interesting and defiant fact is that VR can simulate, not only existing situations, but situations that allow us to obtain data that otherwise would be impossible or improbable, for situational, economical or ethical reasons (Gutiérrez, 2002).

Using VR, psychological interventions can be repeated and prolonged, providing users safe and more attractive, intuitive and naturalistic interfaces that are possible via the integration of real-time computer graphics and sensory input devices that allow users different levels of immersion in the simulated world (Rizzo et al., 2003; Pelissolo et al., 2012), offering researchers the reduction of distracters that exist in real life (Holden, 2005; Brundage and Hancock, 2015).

However, Lu et al. (2011) suggested that virtual environments are not a simple substitution of traditional techniques of intervention, but they can scaffold and augment traditional treatments or experimental situations. In Rizzo et al. (2003) words, virtual reality '*offers the potential to deliver systematic human testing, training and treatment environments, that allow the precise control of complex dynamic 3D stimulus presentations, within which sophisticated behavioral recording is possible*' (p. 244–245). According to them, virtual environments are the alternative for conducting research in an ecologically (more naturalistic) manner with experimental control–control that can get lost when scientists go to the real settings.

As clinical psychologists, researchers have taken advantage of virtual reality to study phenomena, such as social influence (Loomis et al., 1999; Slater et al., 2000; Blascovich et al., 2002; Hoyt et al., 2003; Bailenson and Yee, 2005), context conditioning, behavioral avoidance (Grillon et al., 2006), and social interactions in collaborative environments (Peña et al., 2015a; Peña et al., 2015b; Rangel and Peña, 2016), it was found that these kinds of situations increase realism without diminishing experimental control and expanding the methodological tools that researchers can use (Blascovich et al., 2002).

In this sense, Loomis et al. (1999) mentioned that immersive virtual environments increase the power of experimental research, as they provide the control that research needs, allowing an increase in the situation's realism, so the impact of experimental manipulations is stronger than the impact reached by traditional laboratory situations. At the same time, the use of this kind of technology proves a big source of new data that can be automatically available for their analysis.

In the next section, it will be discussed how the enrichment of treatments and behavioral study scenarios, through the tools that VR offers, enable a better understanding of the psychological phenomenon, since it would allow 'extending' the boundaries of the office walls or the laboratories, allowing individuals under analysis to behave in an increasingly similar way, as they do in everyday life. It will be analyzed how, by incorporating elements in situations of treatment or experimental analysis of behavior, in a controlled manner, VR allows the deployment of increasingly complex behaviors—those that until now we have not been able to fully access as professionals.

3. Virtual Reality Psychological Approach

Pérez Martínez (2011) pointed out that '... *VR comprises the man-machine interface... that allows the user to be immersed in a computer-generated graphics simulation to interact in real time...*' (p. 5). This implies identification of different immersion degrees from developments in different technological platforms, in which the virtual reality environment is placed.

From a technical point of view, the degree of immersion in VR is linked to the input/output devices. VR has been typically classified as desktop-based, augmented and immersive VR. In desktop-based VR, the user can interact with both the real world and the virtual world at the same time. Here the devices needed may be only the keyboard, the monitor, the mouse and/or the game controller. What mainly distinguishes augmented VR is the display device, in which the image can present a direct view of the real world combined with computer graphics. For Immersive Virtual Reality (IVR), the user can respond or interact only with the Virtual Environment (VE) and not at all with the real world. The more common input devices for IVR are the data glove and for the display, the HMD (head-mounted display). Nevertheless, the users' immersion feeling can be attached to other factors, such as their willingness to believe that the virtual environment is real.

Even though the use of VR has been recognized as beneficial in different human activities, including the research area, little has been described about the psychological processes regarding immersion that might be involved in different scenarios in virtual environments.

This, in turn, will drive us to question the changes in which people react to the different elements that determine the situation based on the immersion level to which they have been exposed. These configuration levels can go from the identification of non-immersive virtual environments in which the user is exposed to relate to objects and events in two dimensions (2D) through the computer, to the possibility of generating an environment with a three dimension (3D) configuration, that might imply the total corporal immersion of one or more users in real time with a multi-sensorial feedback.

Furthermore, if it is recognized that the psychological behavior makes reference to the way in which the individual faces the relations established with his/her environment from the type of contact that occurs with such relations (Gibson, 1979; Ribes, 2018). It could identify, as an attempt, different levels of behavioral complexity qualified by different configurations of the organization of the environment; in this case, virtual, faced by the user. According to Ribes (2018), at least five forms of behavioral contact (contingencies) can be recognized and which involve progressive complexity that can help as a reference point to understand the emerging psychological interactions in Virtual Environments (VE):

1. *Coupling contingencies*: In this type of interaction or functional contact, users interact with the elements that configure the environment they face, and in which the organization is independent of their activity. That is to say, the user cannot affect what is happening in the environment, but he/she can be affected by it.

2. *Alteration contingencies*: In this level, the user's behavior is capable of altering the environment organization by manipulating the elements that configure the environment and that can be operated. In this way, the user can operate the virtual environment regarding objects and/or alter other users' activities.

3. *Comparison contingencies*: In this type of contingencies, the functional contact rises when the situations faced by the users are characterized by constant change in the structure. Changes can be of two types: those that are linked to the functional properties of the objects that configure the environment (absolute change), and those that depend on the relational properties between the elements that configure the environment, including the user's own response (relational changes). Consequently, the user's behavior supposes a distinction in the constant and changeable elements of the environment he/she is interacting with.

4. *Extension contingencies*: In this type of interaction, users face the present virtual environment in terms of some development criteria from previous contact in similar

circumstances (virtual or real), or from possible circumstances (concurrent and/or future ones) in which they can 'extend' or translate an efficient way of development by coupling, alternation or comparison in the current situation. In other words, the users face the situation in the VE 'as if' it was a different situation.

5. *Transformation contingencies*: Finally, this is the most complex level of possible interaction. In this interaction, users impose the sense and the configuration possibilities of the situation they face. As its name indicates, in this type of interaction, the user is able to transform the faced relations, not in terms of other situations, but by configuring the sense and the possible interactions in the VE.

Each kind of such functional contacts is subordinated to the reactive capability imposed on the individuals by biology, history and/or possible conditions in the environment. We will later show, the type of possible interaction we think will be closely related to the configuration possibilities in the VEs, from the quantity and quality of the participating elements in a given environment and the interactive possibilities that are allowed by them. These definitely have an important role in the way in which individuals establish functional contact with their environment (in a broad sense), that is, in the way individuals behave psychologically.

So, in the next section, different configuration possibilities that can be identified in VEs, their elements, and the type of immersion allowed by them will be described. After that, it will be analyzed how these configurations could be related to different levels of an individual's psychological processes.

4. Configuration Possibilities in VEs

Traditionally, the technology development related to VR has been described by 3D developers. This means that the users face 'virtual' circumstances, not only in a graphical surface (X, Y axes), but also with depth (Z axis) in which objects and events are presented, generating an 'almost physical reality'. However, we can consider that even in low digital environments with just 2D, a user in an informatics environment faces 'virtual' ways of interaction, since he/she interacts with simulated objects and events in a graphical environment. If this is the case, we could describe different immersion grades in respect to the virtual environment in terms of vivid exposition, and contact with objects and events in function with the flexibility with which the elements, participant, objects, and events are structured.

Table 1 reveals different elements that are considered for configuration of an environment with respect to which technologies related to digital environments and VR have been developed.

If we accept that VR is a 'synthetic experience' in which a 'physical reality' is replaced for the user through a computer (Pérez-Martínez, 2011), then we should recognize that this is the case even if we use graphic platforms in 2D or 3D. In fact, the most important sense for the simulation is the users' capacity of altering or not, the elements that configure the environment in which they are exposed. From this, we could describe different types of environments in terms of the level of objects' alterations or their exposed relations in such a way that we could have environments in which elements stay without alteration, while the user is exposed to them. The denominated Fixed Environments are different from those environments in which elements and their relations can change from time to time. The last two types of environments are

Table 1: Configuration elements of the virtual environment.

User	Type of Environment	Environmental Configuration	Type of Immersion
One	2 D	*Fixed*	*Expository*
Multi-	3 D	Dynamic, not based on the user's responses	*Active*
user		Dynamic, based on the user's responses	*Avatar*
			Corporal

considered Dynamic Environments and they can be: (1) Those in which changes do not depend on the user's activity (Independent Dynamic Environment), and (2) environments in which the changes depend on some kind of response or group of responses from the user (Dependent Dynamic Environment). From the above, we can consider that different degrees of immersion should be related to the type of user participation with the elements of the environment that are possible through the different types of environments. In this context, the next typology of immersion is recognized:

Expository, in which the user has passive interaction with the objects of the environment.

Active, in which the user can manipulate and/or navigate with respect to the objects in the environment.

Avatar, in which the user is represented as one more element in the graphics environment and has operative possibilities with respect to the rest of the elements.

Corporal, in which through technological devices (gloves, glasses, suits, etc.), the users have the possibility of acting and getting multi-sensorial feedback from the different elements that constitute the virtual environment as a consequence of their own relation with them.

It is important to point out that the type of immersion does not refer to any psychological state or expression. It refers to the conditions of exposure and contact with the elements that configure the virtual environment and the possible participation or not of other users. In the next section, we will try to explore the possible psychological relationships that could emerge when each of these independent elements is combined.

5. Psychological Processes and Virtual Reality

Even though the type of configuration of the virtual environment components does not determine *per se* the kind of functional contact, it has to be considered that they could have an important weight in the structuring and emerging of the different interactive systems. They could be considered as factors that create conditions that could propitiate or hinder specific ways of psychological behavior. As aforementioned, within the experimental tradition in the behavior analysis, a number of conditions have been established to evaluate the development with some degree of immersion for the participants, when they are exposed to tasks in which they manipulate different learning conditions (Staddon and Cerutti, 2003). However, not all the immersion conditions proposed here have been evaluated, and the use of 3D graphical environments has not been exploited. Maybe, that is why, from the psychological perspective, the success is less than that expected in the attempt to account for the conditions that would make easier for the individuals to establish complex modes of functional contact by extension or by transformation. Usually, in these types of studies the behavioral interactions are identified and in which the adjustment criteria of the experimental task ask for user adjustment by coupling, by alteration, or by comparison (Peña Correal et al., 2009). And maybe these results are due to the fact that in these used tasks, the participants have contact only with the feedback in a unisensory or bisensory way and which is typically visual and/or auditory and does not necessarily promote complex interactions.

According to Gibson (1979), the movement within an environment in which the individual is immersed, is a powerful adjustment indicator of the extent from the inherent changes in positions to an entity in movement. It promotes the identification of different structural possibilities from environmental arrangement with different textures and estimative proprieties related to individual activity. As a consequence, the environment becomes highly dynamic and the regularities in the relational properties are perceived only through movement.

Linked processes with the functional contacts by extension and transformation imply a high users' participation because the sense of their interactions is not defined by the particular elements that underlay each situation. These types of behavior demand that the environment acquires different possibilities related, not only to the elements that constitute it, but also with the dimensions (sensorial and linguistics) with which the user has contact and participates in its construction and its occurrence. We think that the use of digital platforms in virtual environments with corporal immersion could be propitiatory conditions for the emergence of this type of complex behavior, insofar as they allow a 'real experience' situation, as also the identification of the relevant elements analyzed in different perspectives, by incorporating the movement possibility in the used application with this aim.

In the construction of the VR platform, the movements that users are allowed to make have been implemented in two levels: (1) in the first person, when the user cannot see her/himself interacting through her/his graphical representation, and (2) in the third person, when the user can see her/his graphical representation interacting with the environment and others. It is possible that the perspective that allows this type of manipulation might also influence the degree of immersion feeling in the type of interaction process that the user establishes with the virtual environment.

6. Final Considerations

The aim of this chapter has been to describe how VR is becoming a tool for behavior analysis. Since it allows the configuration of a variety of VEs that do not determine the individual's functional contact, but definitely they could be playing an important role in the way in which psychological behaviors emerge and structure. In this sense, the advantages that VR brings to the arrangement of controlled environments, complex environments overall, are unquestionable and professionally, expand the possibilities to acquire, through this tool, more knowledge about relevant variables involved in psychological behavior.

Nevertheless, in spite of the countless advantages of using virtual reality in behavioral sciences as above mentioned, some authors have also highlighted some disadvantages. One of these is the infinite possibilities that it offers to researchers who actually have simulated a variety of situations that cannot be easily replicated; also the comparison of results and the generated products are complicated (Gutiérrez, 2002). According to Lu et al. (2011), another disadvantage is that it is not an easy task to create the desired sense of reality that the researchers might want for their study. Third, due to some kind of technology that has been developed relatively recently, it is not economically accessible to most researchers, and in most of the cases, they are not familiar with novel applications (Rizzo et al., 2003). Fourth, the imperfections associated with hardware or with modeling software can be difficult to find or to solve for a common researcher as immersive virtual environments demand a considerable development of technological skills (Loomis et al., 1999). However, these disadvantages have to be solved; the rearrangement of close experimental conditions can be made at a low economic cost and in a very short period of time.

In any case, despite these disadvantages, the number of possibilities that virtual reality offers to behavioral analysts and to researchers with different approaches is highly attractive. VR surely will support the advance in questions and answers that will shorten the gap in the complex functional environment with which people interact in real life.

References

Bailenson, J.N. and Yee, N. (2005). Digital chameleons. Automatic assimilation of nonverbal gestures in immersive virtual environments. Psychological Science, 16(10): 814–819. DOI: 10.1111%2Fj.1467-9280.2005.01619.x.

Blascovich, J., Loomis, J., Beall, A.C., Swinth, R., Hoyt, C.L. and Bailenson, J.N. (2002). Immersive virtual environment technology: Just another methodological tool for social psychology. Psychological Inquiry, 13(2): 103–124.

Brundage, S.B. and Hancock, A.B. (2015). Real enough: Using virtual public speaking environments to evoke feelings and behavior targeted in stuttering assessment and treatment. American Journal of Speech-Language Pathology, 24: 139–149.

Calogiuri, G., Litleskare, S., Fagerheim, K.A., Rydgren, T.L., Brambila, E. and Thuston, M. (2018). Experiencing nature through immersive virtual environments: Environmental perceptions, physical engagement, and affective responses during simulated nature walk. Frontiers in Psychology, 8: 1–14. DOI: 10.3389/fpsyg.2017.02321.

Cesa, G.L. Manzoni, G.M., Baccheta, M., Castelnovo, G., Conti, S., Gaggioli, A., Mantovni, F., Molinari, E., Cárdenas-López, G. and Riva, G. (2013). Virtual reality for enhancing the cognitive behavioral treatment of obesity with binge eating disorder: Randomized controlled study with one-year follow up. Journal of Medical Internet Research, 15(6): e113. DOI: 10.2196/jmir.2441.

Dos Santos, A., Borloti, E. and Bender Haydu, V. (2018). Terapia com exposição à realidade virtual e avaliação funcional para fobia de dirigir: um programa de intervenção. Avances en PsicologíaLatinoamericana, 36(2): 235–241.

Ellis, S. (1995). Origins and Elements of Virtual Environments. New York: Oxford University Press.

García García, E.S, Rosa-Alcázar, A. and Olivares, P.J. (2011). Virtual reality exposure therapy and internet in social anxiety disorder: A review. Terapia psicológica, 29(2): 233–243. https://dx.doi.org/10.4067/S0718-48082011000200010.

Gibson, J.J. (1979). The Ecological Approach to Visual Perception. New York: Psychology Press, Taylor & Francis Group.

Goldsmith, T.R. and Le Blanc, L.A. (2004). Use of technology in interventions for children with autism. JEIBI, 1(2): 166–178.

Grillon, C., Baas, J.M.P., Cornwell, B. and Johson, L. (2006). Context conditioning and behavioral avoidance in a virtual reality environment: Effect of predictability. Biological Psychiatry, 60(7): 752.759. DOI:10.1016/j.biopsych.2006.03.072.

Gutiérrez, J. (2002). Aplicaciones de la realidad virtual en psicología clínica. Aula Médica Psiquiatría, 4(2): 92–126.

Holden, M.K. (2005). Virtual environments for motor rehabilitation. Cyber Psychology & Behavior, 8: 187–219. DOI: 10.1089/cpb.2005.8.187.

Hoyt, C., Blascovich, J. and Swinth, K.R. (2003). Social inhibition in immersive virtual environments. Presence Teleoperators & Virtual Environments, 12(2): 183–195. DOI: 10.1162/105474603321640932.

Loomis, J., Blascovich, J. and Beall, A.C. (1999). Immersive virtual environment as a basic research tool in psychology. Behavior Research and Methods, Instruments & Computers, 31(4): 557–564. DOI: 10.3758/BF03200735.

Lu, S., Harter, D. and Pierce, D. (2011). Potentials and challenges of using virtual environments in psychotherapy. Annals of Psychotherapy & Integrative Health, 1: 56–66.

McKenney, A., Dattilo, J., Cory, L. and Williams, R. (2004). Effects of a computerized therapeutic recreation program on knowledge of social skills of male youth with emotional and behavioral disorders. Annual in Therapeutic, XIII: 12–23.

Pelissolo, A., Zaoui, M., Aguayo, G., Yao, S.N., Roche, S., Ecochard, R., Gueyffier, F., Pull, C., Berthoz, A., Jouvent, R. and Cottraux, J. (2012). Virtual reality exposure therapy versus cognitive behavior therapy for panic disorder with agoraphobia: A randomized comparison study. Journal of Cyber Therapy & Rehabilitation, 5(1): 35–43.

Peña Correal, T., Ordoñez, S., Fonseca, J. and Fonseca, L.C. (2009). La investigación empírica de la función sustitutiva referencial. pp. 35–100. En: Padilla Vargas, M.A. and Pérez-Almonacid, R. (eds.). La función sustitutiva referencial: Análisis histórico-crítico/Avances y perspectivas. USA: University Press of the South.

Peña Pérez Negrón, A., Rangel, N. and Lara, G. (2015a). Non-verbal interaction contextualized in collaborative virtual environments. Journal on Multimodal User Interface, 9(3): 253–260. DOI: 10.1007/s12193-015-0193-4.

Peña Pérez Negrón, A., Rangel, N. and Maciel, O. (2015b). Comparison of the effects of oral and written communication on the performance of cooperative tasks. International Review of Social Sciences, 3(11): 487–499.

Pérez Martínez, F.J. (2011). Presente y futuro de la tecnología de la realidad virtual. Revista Creatividad y Sociedad, XVI16: 1–39.

Ploog, B.O., Scharf, A., Nelson, D. and Brooks, P.J. (2013). Use of computer-assisted technologies (CAT) to enhance social, communicative, and language development in children with autism spectrum disorders. J. Autism Dev. Disord., 43: 301–322. DOI: 10.1007/s10803-012-1571-3.

Powers, M.B. and Emmelkamp, P.M.G. (2008). Virtual reality exposure therapy for anxiety disorders: A meta-analysis. Journal of Anxiety Disorders, 22(3): 561–569.

Rangel, N. and Peña-Pérez Negrón, A. (2016). Efectos de la interacción verbal en el desempeño de tareas cooperativas. Revista de Psicología, 25(1): 1–19. DOI: 10.5354/0719.0581.2016.40628.

Ribes, E. (2018). El estudio científico de la conducta individual: Una introducción a la teoría de la psicología. México: Manual Moderno.

Riva, G. (2011). The key to unlocking the virtual body: virtual reality in treatment of obesity and eating disorders. Journal of Diabetes Science and Technology, 5(2): 283–292. DOI: 10.1177/193229681100500213.

Rizzo, A.S., Schultheis, M.T. and Rothbaum, B.O. (2003). Ethical issues for the use of virtual reality in the psychological sciences. pp. 243–280. In: Bush, S.S., Drexler, M.L. and Lisse, N.L. (eds.). Ethical Issues in Clinical Neuropsychology. USA: Swets & Zeitlinger Publishers.

Slater, M., Sadagic, A., Usoh, M. and Schroeder, R. (2000). Small group behavior in a virtual and real environment: A comparative study. Presence Teleoperators & Virtual Environments, 1(9): 37–51. DOI: 10.1162/105474600566600.

Staddon, J. and Cerutti, D. (2003) Operant conditioning. Annual Review of Psychology, 54: 115–144.

Stichter, J.P., Laffey, J., Galyen, K. and Herzog, M. (2014). Social: Delivering the social competence intervention for adolescents (SCI-A) in a 3D virtual learning environment for youth with high functioning autism. J. Autism Dev. Disord., 44: 417–430. DOI 10.1007/s10803-013-1881-0.

Chapter **2**

Navigation in Virtual Reality

Adriana Peña Pérez Negrón,* Graciela Lara López
and *Elsa Estrada Guzman*

1. Introduction

Navigation in virtual reality (VR) is a broad area of research. This technology might
be the key to understand a number of aspects related to navigation skills because VR
proportionate a 3D space in a controlled situation with the possibility of collecting, in
an automatic mode, all kinds of users' interactions with the virtual environment (VE).
However, it is still an open issue to understand the proper design in which virtual reality
can be a proxy of real life for spatial knowledge training. Also, supporting navigation in
virtual environments has not been completely accomplished. New technologies allow
better input/output devices that will constantly deliver advantages over the old ones. In
this chapter, we present an overview through a literary review of navigation in virtual
reality from both approaches—understanding real life navigation in order to support
navigation in virtual environments, and understanding virtual environmental navigation
in order to support real life navigation skill training.

We constantly move from one place to another to relocate ourselves, and somehow
we know in advance the route to follow. Where to go implies to know where we are
and the relative location of significant elements, along with the information update on
this knowledge as we move. This is our navigation skill that compels the abilities of
having sense of direction, location and orientation when moving around (Wolbers and
Hegarty, 2010), where orientation supports a successful navigation, particularly when the
destination is not visible (Parush and Berman, 2004).

Although a common and regular task in our daily activities, navigation is a complex
function with a number of factors involved in navigation functionality. According to
Bouwmeester (2017), navigation integrates many cognitive functions that shape our
spatial cognition. This is a cognitive approach which is most prevalent in navigation
studies.

Universidad de Guadalajara, Mexico.
* Corresponding author: adriana.pena@cucei.udg.mx

Back in the late 40s, Tolman (1948), in order to explain our spatial behavior based on the creation of spatial representations, introduced the term 'cognitive map' to denote a person's internal map. Following this assumption, if the environment to be navigated is known, we use that internal representation, or we can also create the internal map by navigating on the environment. On the other hand, when the environment is not known, we can use external supports, like maps or external cues, such as sunlight or sounds (Boccia et al., 2014). Therefore, when a person navigates he/she acquires information to be used at that time or in the future.

Navigation can be either active or passive. Passive navigation usually involves the presentation of visual information on a path or environment to a static observer, while active navigation requires actual moving around the environment. Active navigation helps to extract information from the environment by combining cognitive and locomotion elements (Darken and Peterson, 2001).

The locomotion element conveys a sensory experience. Faber (2014) differentiated three sensory inputs for navigation: (1) visual information; (2) podo-kinetic information; and (3) vestibular information. Visual information includes seeing changes related to a relative position of landmarks and the optic flow during movement. The sensory input of podo-kinetic information can be described as a collaboration of motor commands and proprioceptive information. It entails knowing where and how one's body parts are located and what muscle commands have been executed to move them. The vestibular information collects information about motion and balance from the head movements.

In the cognitive element, visual-spatial memory constitutes an important factor. This is the process by which we keep spatial or visual information over a limited period of time (Voyer et al., 2017) and that involves the ability to recognize and recall a number of characteristics of the environment and the objects on it, such as shapes, colors, object location and path movements (Liang et al., 2018), as a resource to find our way within that environment, that is, wayfinding.

Navigation can then be divided into locomotion (moving in the environment), and wayfinding (the spatial decision-making) (Satalich, 1995; Montello, 2005). Both terms refer to the purpose of moving though a large-scale environment while maintaining orientation. However, wayfinding requires a goal, a destination or a target for the navigation (Montello and Sas, 2006).

A number of clues help people to find their way; among them is our sense of orientation. Our orientation mechanisms affect the process of determining and following a path. Meilinger (2008) defined orientation as trying to regain one's position. Orientation helps us to know where we are, and our wayfinding makes us know how we can get from there to a certain destination (Dijkstra, 2014).

Back in the 60s, Lynch (1960), an urban planner, described spatial knowledge as a process that starts with landmarks, salient objects; then, routes (or paths) that connect the landmarks, not necessarily directly moving through; and then, nodes that are interchanges or junctions between routes. These structures guide us to an efficient navigation (Darken, 2001). Based on this model can be distinguished: landmark navigation, recognizing certain locations based on landmark knowledge; location navigation by linking landmarks related to route knowledge; and path navigation based on the use of paths for orientation and related to survey knowledge, required to create new paths, shortcuts or for navigation in changing environments (Bouwmeester, 2017). In contrast to orientation, path integration does not involve the recognition of external features, such as geometry or landmarks (Montello, 2005); instead our sensory inputs based on locomotion help us keep track of certain locations in the environment (Meilinger, 2008).

Figure 1 depicts the navigation requirements and relations to locomotion and cognitive elements of active and passive navigation. The cognitive element results on spatial knowledge obtained through navigation experience.

In the effort of the research community to understand how aspects of our physical environment contribute to our spatial cognition, several studies have been conducted by taking advantage of Virtual Reality (VR), mainly because the use of the real world setting for the study of spatial cognition is challenging due to the lack of control and time exposure to the environment. Therefore, using VR as an alternative provides similar to real-world experiences, but with full experimental control (Hasting, 2013). Various studies have proved the benefits of simulated navigation tasks as predictors of real-world functioning (Courtney et al., 2013). However, it should be kept in mind that while research has proved that people could acquire spatial knowledge from virtual environments (VE), the rate of learning and accuracy of performance is almost always inferior to a real-world performance (Lessels, 2005).

And the other way around, understanding people's navigation in real life should be helpful to comprehend how to support navigation in VEs. As Darken and Peterson (2001) pointed out, people are used to navigating in physical spaces, though this is not a reason to copy the real world. We need to understand how people relate to the physical world in order to support navigation in VEs.

A number of aspects differentiate navigation in real life from navigation in VEs, starting with the fact that in VR, navigation does not include physical restrictions; people can, for example, fly or be tele-transported.

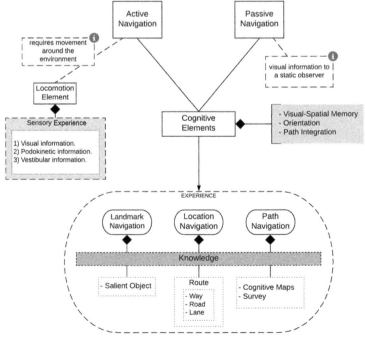

Fig. 1: Navigation elements' relations.

In the next section navigation is first discussed in terms of virtual environments. Then a literary review is presented with papers classified as—those related to navigation aids (Section 3) and those in which navigation is used to study spatial knowledge or for spatial training (Section 4). Finally, Section 5 presents the conclusions.

2. Navigation in Virtual Environments

A starting point would be to distinguish the kind of VR where navigation takes place. Virtual reality systems can be broadly divided into three types (Kalawsky, 1996): non-immersive, also known as desktop systems, semi-immersive and immersive systems.

1. Non-immersive VR is most commonly used today, with a field view of around 50°. In desktop-based VR, the user can interact with both the real and the virtual world at the same time. The only devices required might probably be the keyboard, the monitor and the mouse or a game controller.

2. Immersive VR systems, with a field of view of 360°, represent a technology that excludes the real world around the user. Therefore, the user can interact exclusively with the virtual world. The more common input devices here are a data glove as input device, and for the display, an HMD (head-mounted display) or the CAVE™, a theater 10'X10'X10' made up of three rear-projection screens for walls, and a down projection screen for the floor (Cruz-Neira et al., 1993) as the system shown in Fig. 2.

3. Semi-immersive VR is something between the two previous systems, with a field view ranging typically between 100° to 150°; what mainly distinguishes it is the display device.

Navigation and wayfinding are the most common spatial tasks performed in VR research (Montello, 2005). Navigation in VEs is usually characterized by a user getting around within the environment through the manipulation of a virtual camera, accomplished mainly by walking around or flying over (Liang and Sedig, 2009). This task in desktop-VR will very probably be together with an *avatar* that simulates movement

Fig. 2: A user in the CAVE.

to provide a sense of immersion (Dias et al., 2015), when a first-person *avatar* approach is not used. VR is primarily a visual interface, and it has been found that visual feedback can influence the perception of navigation and even of self-motion (Harris et al., 2002).

However, navigation in VR produces special difficulties when compared with navigation in real life, such as the lack of fidelity and the diminished number of clues that we can find in the real world. It also has to be kept in mind that, while in the real world we can use a number of clues that support navigation along with the sight, like the smell and hearing senses, or for example the sunlight, in VE, the user can mainly rely on the sight and the hearing only when the VE design includes acoustics (Hasting, 2013).

Visual fidelity is the degree to which visual features of the VE match with the visual features of the real world (Hasting, 2013). For computer graphics, the degree of visual fidelity can be classified as (Ferwerda, 2003):

1. Physical realism: Images with the same visual stimulation as the scene; an accurate point-by-point representation with accurate shapes, materials and illumination properties.

2. Photo-realism: Images with the same visual response as the scene; i.e., the rendered image is indistinguishable from a photograph of the scene.

3. Functional realism: Images with the same visual information as the scene; i.e., when the rendered image does the task as well as in the real world.

Due to technical issues, the most commonly used visual fidelity is functional realism. A few studies have looked at the impact of visual realism on navigation (Hasting, 2013). Although, in general, results show that low-fidelity visualizations are more beneficial for tasks that require a cognitive map or survey knowledge, and high-fidelity representations are better in the construction of egocentric spatial representations or route knowledge (Wallet et al., 2011).

As mentioned earlier, proprioceptive feedback allows us to know the position and orientation of our limbs and head, and vestibular feedback gives us a sense of translation and rotation in space; lack of these kinds of locomotion feedback is difficult in navigation in VEs, mainly because of the loss of spatial updating abilities and path integration (Christou et al., 2016).

Inconsistencies in spatial information-updating process are one of the factors to cause disorientation in VR (Klatzky et al., 1998). When using HMD technology, people take a longer time to navigate. These inconsistencies in the spatial update with respect to head movements can also cause motion sickness (Klatzky et al., 1998).

Based on the physical space of the VR environment, navigation can be accomplished by walking, or using an input/output computer device; a control device like the mouse. But virtual walking refers to traveling within a VE and this does not necessarily imitates exact physical movements. By using control devices, a change in direction or moving around can be achieved through the mouse or a game controller, for example. The use of such control devices makes it possible to navigate in arbitrarily large virtual environments, using a small physical space and without the need of additional tracking devices (Nabiyouni et al., 2015). Aiding user navigation techniques aim at improving movement efficiency and precision. Contemporary solutions for the user input interface include: mouse, keyboard, data gloves, motion tracking systems, voice commands or vision-based hand gestures, among others (Wojciechowsky, 2013).

In VR, the direction of navigation is usually associated with the user's view of the scenario, that is, the user navigation direction is in front of him/her. Navigation in non-immersive desktop VR requires a control device; the view of the world changes when the user makes a translation and/or rotation within it. But in immersive VR, the user is not restricted to looking in just one direction regarding the physical space. Here the user usually can use head and body rotations to change direction. The possibility of head movements while walking allows gaze-direction motion control (Christou et al., 2016). In summary, navigation in VR can be accomplished by physical movements through a device or by a combination of both based on the design of the VE (Christou et al., 2016).

Bowman (2001) categorized five metaphors in VR for navigation:

1. Physical movement (walking) achieved by using motion trackers and locomotion devices.
2. Manual viewpoint manipulation through a user's hand motion to pull him/herself or object selection to which towards the user moves.
3. Steering; the continuous specification of the direction through gaze-direction or pointing.
4. Target-based travel by specifying a destination (teleportation).
5. Route planning; when the user specifies the path and the system handles the movement.

Likewise, Bowman (2002) presented a motivation-based approach by dividing the navigation process in:

- Exploration—navigation with not specified goal.
- Search—navigation with a searching goal.
- Maneuvering—slight movements to correct position and orientation.

As mentioned earlier, in VR, people might navigate by walking or flying, or they can be tele-transported; that is, to be moved instantaneously from their position to a different location. This generates two types of displacement: (1) continuous, when the viewpoint is moved fluently; and (2) discontinuous, when the viewpoint can be moved instantaneously over arbitrary large distances. Discontinuous displacement makes it difficult to create a cognitive map, although this can be diminished by anticipating the destination (Bakker et al., 2001).

Regarding spatial awareness, the most natural way in which the user receives proprioceptive and vestibular stimuli is, of course, through walking. However, this is the most obstructive one due to physical space restriction, which has been overcome by, for example, treadmills—although a costly solution. Other software-proposed alternative solutions are redirection, resetting and scaling techniques (Chen et al., 2015). In this regard it has been found that by using an HMD, the lack of body rotation or translation generates a worse performance in spatial cognition than walking and physical rotating. But, only physical rotation performance is comparable to providing physical rotation and translation (Ruddle and Lessels, 2006, 2009).

A combination of head-track for rotation with translation using a control seems to be a locomotion method that supports awareness, which can be accomplished by gaze-direction or pointing direction (Christou et al., 2016), see Fig. 3.

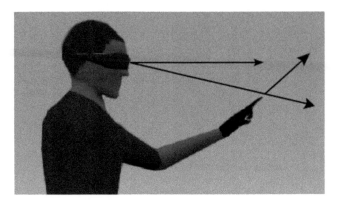

Fig. 3: Gaze-direction or pointing direction.

Summarizing, a number of design considerations in the VE might represent a disadvantage for navigation in VR when compared to real life, such as the lack of fidelity, smell and hearing senses, locomotion feedback, spatial information proper update, and the navigation direction associated with the user's view. Among other VE design considerations for navigation are control velocity and constrains. On the other hand, VR allows the design of aids for navigation that might not be available in a real-world situation. However, how to help users to find their way, or to support proper navigation performance in VEs is still an open issue.

With the intention of getting an overview of the new tendencies for navigation in VE, a search was made in Google Scholar with the chain: 'allintitle': 'virtual reality'|'virtual environment', 'spatial orientation'|'navigation'. We limited the search to papers not older than five years (starting January 2013) and got an outcome of 170 papers. From these papers only those that presented empirical studies were selected. We also excluded those studies aimed at treating a disease or handicapped person. Eighteen primary papers were classified as those for navigation aids based on hardware or software, although some can be categorized in both. In those related to the use of VR to study spatial knowledge or training, primary papers were numbered and distinguished in the paper by placing [P1.. P18], after its citation. In the next section is presented the subject of navigational aids— first an overview and then the selected papers from the literary review.

3. Navigation Aids in VR

3D videogames are VR; they present 3D scenarios where user's interaction or in this case, player's interaction is allowed. Liszio and Masuch's (2016) work is related to videogames, but it can be transposed to VR. According to Liszio and Masuch (2016), supporting player navigation can use the next techniques:

1. Focusing the player's attention on relevant objects or locations in the game world, so that are used as landmarks.

2. Guidance systems, such as signs and arrows, or specific equipments like maps and compasses.

3. Influencing the player's decisions; one way to accomplish this is by constraining navigation.

Liszio and Masuch (2016) also compiled 28 design patterns addressed to identify navigation problems that go from landmarks, interactive objects or maps to ambient sounds.

It seems evident that the user's navigation in VE can be supported, same as in real life, using resources like landmarks or maps. However, it has to be kept in mind that there are other psychological factors that might influence the user's adaptation to a computer-generated environment, such as gender, age, or his/her familiarity with the VE (Christou et al., 2016).

A number of navigational aids in VR are electronic analogues of the tools from the real world; among these the most common is a map of the environment, though in VE features unable to do so in paper map can be added, such as self-orientation and real-time user's position (Burigat and Chittaro, 2007). A route list or instructions can also be used; however, they are less efficient. 3D maps can also be generated within the VE, presenting a miniature of the VE itself (Chittaro et al., 2005).

Most of games with navigation in landscapes provide a map, usually presented in two ways: toggled on/off, covering the screen to interact with the map or a mini-map heads-up display with no interaction that follows the player. With big landscapes, come big maps that cover an extensive part of the screen. Therefore they are often transparent, allowing the user to interact with the map and make a quick change in the game or interact with both at the same time. It has to be kept in mind that when displaying a complex map, it needs to be organized and intuitive (Gierloff, 2014).

There are a number of possibilities in VR that can guide the user's navigation. Guided navigation can be implemented in VEs by, for example, placing arrows or through an animated guide, or moving the user in a vehicle. Furthermore, users can have special powers, such as seeing through occluding surfaces or traveling through them (Burigat and Chittaro, 2007).

Understanding navigation activities will better support wayfinding aids. In VE, wayfinding can be as simple as moving the user to the destination when it is known, but these tools deprive the user from an active navigation (Wu et al., 2009). In order to get a more efficient navigation in self-exploration, constraining the degree of freedom helps the user, but apparently this provokes confusion in large VEs (Ahmed and Eades, 2005). Environmental design elements can also support wayfinding. The use of layout, landmarks and signage helps users, particularly in unfamiliar environments.

Wu et al. (2009) conducted a study with three types of aids: view-in view map, animation guide, and human system collaboration. The different degree of effectiveness was found related to people with different spatial abilities. Therefore, the design of navigation aids should consider the type of spatial information based on the type of task and the spatial ability of the user.

New forms to aid navigation through software are constantly appearing. Next are presented some of the papers from the literary review, related to integrate aids for navigation in VEs. Supporting navigation in VR can be categorized as those that propose approaches through the use of input/output devices, that is, hardware-based approaches, and those that aid navigation through software design within the environment. First are presented those studies that use hardware in their approach to facilitate navigation for the user, and then those that are based on software or environmental design to support navigation. It has to be kept in mind that even if the approach is hardware-based, software or environmental design might be also required for its adaptation.

3.1 Navigation Aids Based on Hardware for VR

Contextualized in urban planning and building design, where the user sees the model from a ground view, Roupé et al. (2014) [P1] proposed the use of poses and gestures for navigation in desktop-VR by using the XBOX Kinect™. The used body postures are: a leaning-forward body of 10 degrees triggers acceleration; a leaning-backward body of 7 degrees triggers the backward acceleration, a 10 degree rotation of the shoulders triggers a left or right turn, and lifting the right arm fast forward stops the walking. According to their study, most of the participants found this approach easy and not demanding when compared with the use of the keyboard and mouse as the navigation interface. Although, participants expressed that a faster walking speed and a greater degree of freedom to manipulate the view could improve the approach.

Dias et al. (2015) presented two methods to navigate in large display VR (semi-immersive VR) using gestures detected through a depth sensor; they also used Microsoft Kinect™ as input device. In order to improve usability, the authors used natural and simple gestures easy to learn. The first method called 'Bike' is a control that uses the relative position of the hands. The user put his/her hands, emulating the control of a bicycle and placing both hands alongside with closed fists, like when riding a bike. Then, when he/she puts the right hand forward and the left hand back, the camera turns left and vice versa to turn right. To increase speed, the user can step forward or backward to decrease it. The second method called 'free hand' is a control similar to the browse movement of the dominant hand and also similar to the mouse-based interface. Control of the view is made by rotating the dominant hand placed with the palm in front of the display. Navigation speed increases when the user steps forward. Users performed better with the second method. The main constraint on the first method was that the user could not stop the interaction efficiently.

Christou et al. (2016) conducted an empirical study to compare gaze-direction and pointing direction for motion control in a CAVE, using a wayfinding task. The task comprised finding a target-building indicated on a map, with only one distinct route connection to four different starting points. Subjects verbally expressed their preference for the pointing method. The study results suggested that uncoupling the gaze direction from the traveling direction improves navigation performance.

Another approach using the Microsoft Kinect™ to identify gestures was proposed by Vultur et al. (2016). They developed an algorithm for gesture recognition in navigation in VE, based on seven gestures made with the user's body moving forward, backward, to the right, to the left, up, down, and forward with higher speed. In their study, the accuracy rate of gesture recognition with five users was in the range of 87 per cent (in downhill gesture) to 95 per cent (in away gesture). Users expressed a positive experience from the trial as a navigation.

Ragan et al. (2017) proposed an amplified head rotation in VEs particularly for training, when using head movements to view control. As they mentioned, this approach presents an unrealistic mapping, which might cause disorientation. Ragan et al. (2017) conducted an experiment to investigate the effects of this approach on spatial orientation and the performance of a search task in a CAVE with HMD technology. They found that differences in the visible range and display type influence effectiveness and usability of the approach. For orientation, better results were obtained in the CAVE than with the HMD, in which users experienced adaptation troubles.

In mobile VR HMD navigation, Park and Lee (2018) [P6] proposed the use of hand gestures. This approach is based on the drone-flying principle, user-centric navigation, and applying the decoupled degree of freedom (DOF) of navigation from the DOF of visualization to support head and body movement. The hand gesture inclination determines the navigation direction and speed; the tilt and the movement of the palm simulate the drone movements. They conducted comparative studies in navigation tasks to evaluate a button and inertial sensor-based, the most widely used interfaces, in mobile VR, to contrast their hand gesture interfaces in mobile HMD with the keyboard in desktop VR. Results showed that the metaphor for mobile VR is most suitable, while in desktop, VR users faced difficulties in rotating their bodies when seated.

Six studies based on hardware were found and in which a recurrent topic is the use of different gestures for navigation. Three of them used the Microsoft Kinect™, with hand gesture detection or body poses, and one more proposed a different hand-gesture navigation although with a sensor-based technology is a study to compare head and pointing-direction navigation, with results that suggested decoupled gaze-direction from navigation releasing the users FOV (field of view) from the navigation direction. [P5] proposed an amplified head rotation to get a better view of the environment.

3.2 *Navigation Based on Software or Environmental Design in VR*

Terziman et al. (2013) introduced a new approach to improve the sensations related to locomotion in VE. A traditional camera motion typically recreates the human-walking visual flow by oscillation. In this approach, the camera motions include: (1) multistate, that is, locomotion states for walking, running and sprinting; (2) if personified, it can be adapted to the *avatar*'s physiology, such as size, weight or training status; and (3) it considers the topography of the virtual terrain. In their study, the participants could discriminate between the different locomotion modes and some properties, like the *avatar* training status and age.

Chen et al. (2015) presented a collaborative task with co-located people in, for example, a CAVE, a joystick-based navigation that avoids collaborator and/or walls collisions. They integrated a distance limitation considering translation and rotation. To avoid collisions and occlusion among users, they suggested assigning a safe zone for the user and a shared zone for closely coupled interactions. In the shared zone, they adjust a border that will depend on the penetration of other users in the shared area. They conducted an empirical study with results showing that using adaptive workspace boundaries and neutral direction reorientation presented the best cohabitation performance for users' collision and occlusion.

For networked virtual environments (NVE), particularly large streaming VE, Forgione et al. (2016) proposed bookmark aids for the user to move quickly from one position to another by a mouse click. A bookmark is a camera with a predefined position, and it can be presented to users in different ways: as a text link (URL), a thumbnail image, or a 3D object embedded within the NVE itself. This is a type of teleportation with different points of interest, particularly helpful to overcome the latency produced in streaming VEs. They made a study to understand how helpful the bookmarks are for the users through a questionnaire, in which participants found the task achievement easier when using the bookmarks. They also present the good impact on streaming by using pre-computations derived from the use of bookmark points.

Nguyen-Vo et al. (2017) made an interesting proposal for virtual environment with no landmarks. Their proposal used visual overlaid wire frames of a rectangular box as reference. They conducted an empirical study, using a search task for two approaches of the visual frames: an egocentric frame, centered on the user and simulating a CAVE; and an allocentric frame, environment-centered, that simulates a room. Results showed that the allocentric, alongwith being the preferred one by the users, was helpful in the navigational search task.

A number of studies have been conducted to study the influence of proprioceptive and vestibular information for the user, as will be commented in the next section. However, the approach included in this section contains adaptive situations. In this section, it is also the only paper that approaches navigation from a collaborative point of view [P8], constraining navigation to avoid co-users collisions. In [P9] the main aim is to support streaming for networked VEs by using bookmarks, which in turn support navigation by teleportation. And in [P10], the problem of VE with no landmarks is tackled.

The next section presents the use of VR as a means for navigation skill training, including conditions that better support spatial knowledge acquisition.

4. Spatial Knowledge Training in VR

Empirical studies in spatial condition are frequently conducted by using VR (Grübel et al., 2017). VR is an inherent spatial technology and thus a promising medium for training in spatial abilities (Lee et al., 2009). Spatial knowledge acquired in VR has been found to be successfully transferred to the real world in such a way that exposure to a virtual training environment should facilitate spatial knowledge (Wu et al., 2009). Performance in navigation tasks is the most typical approach used to measure spatial knowledge acquisition.

However, while some studies present no difference by using VEs when compared to real world, for example Hahm et al. (2006) conducted a study showing that active navigation is more effective than passive navigation for route learning in both conditions of VR and real world. Other studies showed differences, for example, regarding age. Sayers (2004) made a study in which he found that in VEs, people older than 45 years old got worse results in achieving tasks, like finding objects, than the younger people. As the author pointed out, this not necessarily means that older people are worse at orientation, but they are less familiarized with VEs.

Other differences in studies using VR can be found regarding the gender. Emil Bergvist (2015) found that men got better results in spatial tasks, such as finding objects. In this study, men, as compared to women, showed better spatial orientation, latency, path length, and better management of spatial anxiety. An important factor worth mentioning is the fact that men had more experience in videogames and VEs. Though women felt more anxious regarding orientations and navigation skills in VEs, they were able to better describe the size, the age, color, type of residence, objects and the lighting in the environment.

Likewise, technical and design conditions of the VR scenario can interfere with the training conditions. For example, studies do not agree on the performance benefits for spatial knowledge in different immersive display variables. This might be due to the diversity in displays, but also with the design condition of the studies, like the task measures, performed tasks, variances between individuals, and the presented stimuli.

Consequently, it is not easy to determine if the different immersive displays have an impact on the performance of navigation tasks in VEs (Hasting, 2013).

Furthermore, Hasting (2013) questioned whether the brain performs tasks in VR the same way that it performs in naturalistic conditions. According to them, although people orient themselves correctly and are able to perform virtual navigation tasks in VEs, we use functional imaging instead of motor, proprioceptive and vestibular information in this condition.

There are a number of studies that integrate locomotion feedback to understand its implications. Next are briefly presented some of the primary papers found in the literature review regarding the use of VR in empirical studies related to training for spatial knowledge or navigation skills.

4.1 Trends in Spatial Knowledge Training in VR

Courtney et al. (2013) used VR to study the implications of threat as a distraction from focus on landmark and route knowledge. Participants were exposed to the route-learning task that consisted of following a guide through six zones with different degrees of threat level, followed by a navigation task in which the participants were asked to return to the starting point of the tour. The goal of the study was to get better strategies for systems development with psycho-physiological computing adapted to individuals in order to foster learning settings. They used artificial neural networking to predict the performance outcomes which could result in, as the authors claim, informing adaptive systems design in this regard.

Wallet et al. (2013) conducted a study with 64 participants (half of them being men and half, women) to examine the effect of passive navigation versus active navigation in virtual and real environments in order to understand spatial learning. Spatial learning consists of on-route learning in a virtual district under four conditions of navigation: passive/ground, passive/aerial, active/ground, or active/aerial and which were evaluated for three types of tasks: wayfinding, sketch mapping, and picture-sorting. Results for the wayfinding task showed that participants at the ground-level viewpoint performed better than those at the aerial level viewpoint, especially in active navigation. In the sketch-mapping task, aerial-level learning showed better performance than the ground-level condition, while active navigation was only beneficial in the ground-level condition. The best performance in the picture-sorting task was obtained with the ground-level viewpoint, especially with active navigation. This study confirmed the benefit of active navigation linked to egocentric frame-based situations.

Landmarks are a powerful reference for users' navigation wen Ploran et al. (2014) conducted a study to examine the relation between visual scanning in a not known environment and the accuracy in navigation to a previous learned destination and the accuracy of the mental representation of the environment. They quantified the movements for visual scanning and as expected, successful navigation in a new environment relies on deliberated visual scanning to locate landmarks and cues to guide to the target location. Two unexpected results related to the amount of visual scanning were: the use of the camera for scanning was predictive of worse navigation accuracy. It seems that participants that relied on the camera were less likely to internalize the sequence of movements to a target location; and spending more time in view mode did not predict greater accuracy in remembering the relation of landmarks.

In his master thesis, Faber (2014) studied how valid is VR as a research method for navigation studies by comparing navigation with and without locomotion. The route memory performance was collected under four conditions: real world, desktop VR, desktop VR with a compass aid, and a hybrid condition with locomotion for a virtual route. The participants got extra information from a mobile device. Results showed that the performance in desktop VR with no podokinetic and vestibular stimuli was worse than real world, while the hybrid condition yielded similar results than the real-world condition.

In other master theses, according to studies conducted by Singurdarson (2014) with desktop-VR and HMD, rotation cues did not improve spatial orientation performance, but structured visuals did. It seems that due to lack of physical motion, under certain conditions, spatial updating can be triggered through visual cues (Riecke et al., 2007). In his study, Singurdarson (2014) combined rotation and translation to yield a smoothly curved path similar to naturalistic motion.

In this regard, Plouzeau et al. (2015) proposed the use of proprioceptive vibrations to allow feeling of imaginary movement to decrease simulator sickness generation and improving the sense of presence. In their study, the participants' muscle spindles were excited by vibrators to simulate the pulses sent to the brain during navigation. The user then felt the movement while remaining in place. Results showed that simulator sickness decreased in about 47 per cent. However, proprioceptive vibrations had no impact on the sense of presence or the navigation performance.

Bouwmeester (2017) conducted a study to understand how VR navigation training affects navigation ability and some of its components. Eighty-six participants were divided into four groups: 23 to a control group, 24 to a control group with psycho-education (about egocentric and allocentric navigation strategies), and 18 to egocentric and 21 to allocentric navigation training strategies groups. To measure participants' navigation skill and navigation subtasks or components skills, the Virtual Tübingen Task was used in pre- and post-tests. The navigation training consisted egocentric and allocentric versions in a virtual environment, using gamification elements. Participants performed subtasks with questions about the route (taken from Claessen et al., 2016): (1) in route sequence, participants had to give the order of eight turns; (2) in route continuation, participants were asked how the route continued from eight different images; (3) in distance estimation, participants saw eight images and two scenes, after which they were asked to decide which image was closest to a given scene; (4) in orientation, the participants were asked to point to the starting or ending position according to the scene presented; (5) and in location, the participants were asked to indicate a scene on a map. Effects of virtual reality navigation training were analyzed in overall navigation ability and the components of the six subtasks. In this study, participants did not improve their navigation abilities with the virtual reality navigation training, either in the overall navigation ability or in any of the six subtasks.

Liang et al. (2018) investigated the effects of collaboration and competition activities on users' navigation behavior and acquisition of spatial knowledge within the HMD-based VR environments. With that aim, they developed a multiuser 3D virtual shopping mall with three floors. Twenty-five participants were allocated into three groups: five dyads in the cooperation group, five dyads in the competition group, and five participants in the single trials. The task consisted of collecting items from the scenario. According to their results, they recommended four design guidelines for the next goals: (1) for maximizing memory recall, a single user performs better in dyads' use in a collaborative context; (2) for minimizing navigation time, the better context is dyads in competitive tasks; (3) for maximizing memory recall of spatial information, it is perhaps better to avoid competitive tasks; and (4) for large VE, exploration dyad in collaboration perform better.

As can be observed, there are no conclusive results. Bouwmeester (2017) compared his results with those presented in Claessen et al. (2016), where results for chronic stroke patients showed improvement in overall navigation ability and in four subtasks, using VR for training. Although there were other differences in healthy people as compared with sick people, particularly on the empirical design, this is an example of the contradictory results that have been found in trying to explain whether VR is a reliable source for training in spatial abilities.

The other open issue is locomotion linked to our spatial updating versus the application of visual cues with the same aim (e.g., Faber, 2014; Singurdarson, 2014; Plouzeau et al., 2015) as to which type of locomotion feedback and/or the type of visual cues, and under which conditions these approaches better support spatial updating require further studies.

In any case, VR provides not only a control environment for spatial studies, but also a support for prediction in participants' spatial behavior (e.g., Courtney et al., 2013). VR also provides different perspectives for studying spatial abilities that might not be available in real world (e.g., Wallet et al., 2013).

5. Conclusions

Virtual reality represents a visual base technology in which navigation can be accomplished in a number of forms and where navigation is influenced by the platform of technology in which the virtual environment is allocated. A fascinating area of study both ways: understanding people's navigation in real life and real-life navigation in order to support navigation in virtual environments.

A number of studies have been presented here, related to navigation in VR. We classified them as those with the aim of aiding navigation in virtual environments based on hardware devices or software and design-based environments, and those with the aim of studying the benefits of training in virtual and to translate that to spatial knowledge for real life.

Researchers are still on the way to determining if VR is a proper approach or rather, under which conditions it is a proper approach to train people in spatial knowledge. Also there is a gap in new technologies, new devices to support navigation and the design of better support that can conduct users to a better performance on spatial tasks.

References

Ahmed, A. and Eades, P. (2005). Automatic camera path generation for graph navigation in 3D. Proceedings of the 2005 Asia-Pacific Symposium on Information Visualisation, 27–32.

Bakker, N.H. (2001). Spatial Orientation in Virtual Environments. Doctoral Dissertation, TU Delft, Delft University of Technology.

Bakker Niels, H., Passenier Peter, O., Werkhoven Peter, J., Henk G. Stassen and Peter A. Wieringa. (2001). Navigation in Virtual Environments. Deljt University of Technology, OCP Man-Machine Systems, Mekelweg 2: 2628 CD Deljt, The Netherlands.

Bergqvist, E. (2015). Spatial orientation & imagery: What are the gender differences in spatial orientation and mental imaging when navigating a virtual environment with only auditory cues.

Boccia, M., Nemmi, F. and Guariglia, C. (2014). Neuropsychology of environmental navigation in humans: Review and meta-analysis of fMRI studies in healthy participants. Neuropsychological Review, 24: 236–251. doi:10.1007/s11065-014-9247-8.

Bowman, D.A., Kruijff, E., LaViola Jr, J.J. and Poupyrev, I. (2001). An introduction to 3-D user interface design. Presence: Teleoperators & Virtual Environments, 10(1): 96–108.

Bowman, D.A. (2002). Principles for the design of performance-oriented interaction techniques. Handbook of Virtual Environments, 277.

Bouwmeester, J. (2017). The Effects of Virtual Reality Navigation Training in Healthy Individuals (Master's Thesis).

Burrigat, S. and Chittaro, L. (2007). Navigation in 3D virtual environments: Effects of user experience and location-pointing navigation aids. International Journal of Human-Computer Studies, 65(11): 945–958.

Chen, W., Ladeveze, N., Clavel, C., Mestre, D. and Bourdot, P. (2015, March). User cohabitation in multi-stereoscopic immersive virtual environment for individual navigation tasks. pp. 47–54. In: 2015 IEEE Virtual Reality (VR), IEEE.

Chittaro, L., Gatla, V.K. and Venkataraman, S. (2005, November). The interactive 3D breakaway map: A navigation and examination aid for multi-floor 3D worlds. In: Cyberworlds, 2005, International Conference on IEEE.

Christou, C., Tzanavari, A., Herakleous, K. and Poullis, C. (2016, April). Navigation in virtual reality: Comparison of gaze-directed and pointing motion control. pp. 1–6. In: Electrotechnical Conference (MELECON), 2016, 18th Mediterranean, IEEE.

Claessen, M.H., van der Ham, I.J., Jagersma, E. and Visser-Meily, J.M. (2016). Navigation strategy training using virtual reality in six chronic stroke patients: A novel and explorative approach to the rehabilitation of navigation impairment. Neuropsychological Rehabilitation, 26(5-6): 822–846.

Courtney, C.G., Dawson, M.E., Rizzo, A.A., Arizmendi, B.J. and Parsons, T.D. (2013, July). Predicting navigation performance with psychophysiological responses to threat in a virtual environment. pp. 129–138. In: International Conference on Virtual, Augmented and Mixed Reality. Springer, Berlin, Heidelberg.

Cruz-Neira Carolina, Sandin Daniel, J. and DeFanti Thomas, A. (1993 September). Electronic Visualization Laboratory (EVL), The University of Illinois at Chicago, Surround-Screen Projection-Based Virtual Reality: The Design and Implementation of the CAVE.

Darken, R.P. and Peterson, B. (2001). Spatial Orientation, Wayfinding, and Representation. Handbook of Virtual Environment Technology, Stanney, K. (ed.).

Dias, P., Parracho, J., Cardoso, J., Ferreira, B.Q., Ferreira, C. and Santos, B.S. (2011, August). Developing and evaluating two gestural-based virtual environment navigation methods for large displays. pp. 141–151. In: International Conference on Distributed, Ambient, and Pervasive Interactions. Springer, Cham.

Dijkstra, J., de Vries, B. and Jessurun, J. (2014). Wayfinding search strategies and matching familiarity in the built environment through virtual navigation. Transportation Research Procedia, 2: 141–148.

Faber, A.M.E. (2014). Can Navigation Research Employ Virtual Reality Techniques Without Reducing External Validity? (Master's Thesis).

Ferwerda, J.A. (2003). Three varieties of realism in computer graphics. Proceedings SPIE Human Vision and Electronic, 3: 290–297. doi:10.1117/12.473899.

Forgione, T., Carlier, A., Morin, G., Ooi, W.T. and Charvillat, V. (2016, May). Impact of 3D bookmarks on navigation and streaming in a networked virtual environment. In: Proceedings of the 7th International Conference on Multimedia Systems (p. 9), ACM.

Gierløff, H. (2014). Navigation by the Use of Maps in Virtual Reality Portrayed by Oculus Rift (Master's Thesis, Institutt for bygg, anleggog transport).

Grübel Jascha, Thrash Tyler, Ho Christoph, Ischer and Schinazi Victor, R. (September 2017). Evaluation of a Conceptual Framework for Predicting Navigation Performance in Virtual Reality. Department of Humanities, Social and Political Sciences, ETH, Zürich, Switzerland.

Hahm, J., Lee, K., Lim, S.L., Kim, S.Y., Kim, H.T. and Lee, J.H. (2006). A study of active navigation and object recognition in virtual environments. Annual Review of Cyber Therapy and Telemedicine, 4: 67–72.

Harris, L.R., Jenkin, M.R., Zikovitz, D., Redlick, F., Jaekl, P., Jasiobedzka, U.T. and Allison, R.S. (2002). Simulating self-motion I: Cues for the perception of motion. Virtual Reality, 6(2): 75–85.

Hastings, B.L. (2013). The Influence of Shading, Display Size and Individual Differences on Navigation Performance in Virtual Reality in an Applied Industry Setting (Doctoral Dissertation, Communication, Art & Technology: School of Interactive Arts and Technology).

Kalawsky, R.S. (1996). Exploiting Virtual Reality Techniques in Education and Training: Technological Issues, Loughborough University of Technology, Advanced VR Research Centre, Loughborough: Sima.

Klatzky, R.L., Loomis, J.M., Beall, A.C., Chance, S.S. and Golledge, R.G. (1998). Spatial updating of self-position and orientation during real, imagined, and virtual locomotion. Psychological Science, 9(4): 293–298.

Lee, E.A.L., Wong, K.W. and Fung, C.C. (2009). Educational values of virtual reality: The case of spatial ability. Proceedings of World Academy of Science, Engineering and Technology, Paris, France.

Lessels, S. (2005). The Effects of Fidelity on Navigation in Virtual Environments. The University of Leeds, School of Computing.

Liang, H.N. and Seding Kamran. (March 2009). Characterizing Navigation in Interactive Learning Environments.

Liang, H.N., Lu, F., Shi, Y., Nanjappan, V. and Papangelis, K. (2018). Evaluating the effects of collaboration and competition in navigation tasks and spatial knowledge acquisition within virtual reality environments. Future Generation Computer Systems.

Liszio, S. and Masuch, M. (2016, November). Lost in open worlds: design patterns for player navigation in virtual reality games. *In*: Proceedings of the 13th International Conference on Advances in Computer Entertainment Technology, p. 7, ACM.

Lynch, K. (1960). The Image of the City. Cambridge: MIT Press.

Meilinger, T. (2008). MPI Series in Biological Cybernetics No. 22, Max-Planck-Gesellschaft.

Montello, D. (2005). Navigation. pp. 257–294. *In*: Shah, P. and Miyake, A. (eds.). The Cambridge Handbook of Visuospatial Thinking, Cambridge, Cambridge University Press.

Montello, D.R. and Sas, C. (2006). Human factors of wayfinding in navigation. *In*: Karwowski, W. (ed.). International Encyclopedia of Ergonomics and Human Factors. 3(2): 2003–2008.

Nabiyouni, M., Saktheeswaran, A., Bowman, D.A. and Karanth, A. (2015, March). Comparing the performance of natural, semi-natural, and non-natural locomotion techniques in virtual reality. pp. 3–10. *In*: 3D User Interfaces (3DUI), 2015 IEEE Symposium on, IEEE.

Nguyen-Vo, T., Riecke, B.E. and Stuerzlinger, W. (2017, March). Moving in a box: Improving spatial orientation in virtual reality using simulated reference frames. pp. 207–208. *In*: 2017 IEEE Symposium on 3D User Interfaces (3DUI), IEEE.

Park, K.B. and Lee, J.Y. (2018). New design and comparative analysis of smartwatch metaphor-based hand gestures for 3D navigation in mobile virtual reality. Multimedia Tools and Applications, 1–21.

Ploran, E.J., Bevitt, J., Oshiro, J., Parasuraman, R. and Thompson, J.C. (2014). Self-motivated visual scanning predicts flexible navigation in a virtual environment. Frontiers in Human Neuroscience, 7: 892.

Plouzeau, J., Paillot, D., Chardonnet, J.R. and Merienne, F. (2015). Effect of proprioceptive vibrations on simulator sickness during navigation task in virtual environment.

Parush, A. and Berman, D. (2004). Navigation and orientation in 3D user interfaces: The impact of navigation aids and landmarks. International Journal of Human-Computer Studies, Department of Psychology, Carleton University, Industrial Engineering and Management, Technion, Israel Institute of Technology, 61(3): 375–395.

Ragan, E.D., Scerbo, S., Bacim, F. and Bowman, D.A. (2017). Amplified head rotation in virtual reality and the effects on 3D search, training transfer, and spatial orientation. IEEE Transactions on Visualization and Computer Graphics, 23(8): 1880–1895.

Riecke, B.E., Cunningham, D.W. and Bülthoff, H.H. (2007). Spatial updating in virtual reality: The sufficiency of visual information. Psychological Research, 71(3): 298–313. doi:10.1007/s00426-006-0085-z.

Roupé, M., Bosch-Sijtsema, P. and Johansson, M. (2014). Interactive navigation interface for virtual reality using the human body. Computers, Environment and Urban Systems, 43: 42–50.

Ruddle, R.A. and Lessels, S. (2006). For efficient navigational search, humans require full physical movement, but not a rich visual scene. Psychological Science, 17(6): 460–465.

Ruddle, R.A. and Lessels, S. (2009). The benefits of using a walking interface to navigate virtual environments. ACM Transactions on Computer-Human Interaction (TOCHI), 16(1): 5.

Satalich, G. (1995). Navigation and Wayfinding in Virtual Reality: Finding the Proper Tools and Cues to Enhance Navigational Awareness. University of Washington.

Sayers, H. (July 2004). Desktop Virtual Environments: A Study of Navigation and Age, School of Computing and Intelligent Systems. University of Ulster, Magee Campus, Londonderry, BT48 7JL, UK.

Sigurdarson, S. (2014). The Influence of Visual Structure and Physical Motion Cues on Spatial Orientation in a Virtual Reality Point-To-Origin Task (Doctoral Dissertation, Communication, Art & Technology: School of Interactive Arts and Technology).

Terziman, L., Marchal, M., Multon, F., Arnaldi, B. and Lécuyer, A. (2013). Personified and multi-state camera motions for first-person navigation in desktop virtual reality. IEEE Transactions on Visualization and Computer Graphics, 19(4): 652–661.

Tolman, E. (1948). Cognitive maps in rats and men. Psychological Review, 55: 189–208.

Voyer, D., Voyer, S.D. and Saint-Aubin, J. (2017). Sex differences in visual-spatial working memory: A meta-analysis. Psychonomic Bulletin & Review, 24(2): 307–334.

Vultur, O.M., Pentiuc, S.G. and Lupu, V. (2016, May). Real-time gestural interface for navigation in virtual environment. pp. 303–307. *In*: Development and Application Systems (DAS), International Conference on, IEEE.

Wallet, G., Sauzéon, H., Pala, P.A., Larrue, F., Zheng, X. and N'Kaoua, B. (2011). Virtual/real transfer of spatial knowledge: Benefits from visual fidelity provided in a virtual environment and impact of active navigation. Cyberpsychology, Behavior and Social Networking, 14(7-8): 417–23. doi:10.1089/cyber.2009.0187.

Wallet, G., Sauzéon, H., Larrue, F. and N'Kaoua, B. (2013). Virtual/real transfer in a large-scale environment: Impact of active navigation as a function of the viewpoint displacement effect and recall tasks. Advances in Human-Computer Interaction, 2013: 8.

Wu, A., Zhang, W. and Zhang, X. (2009). Evaluation of wayfinding aids in virtual environment. Intl. Journal of Human–Computer Interaction, 25(1): 1–21.

Wojciechowski, A. (2013). Camera navigation support in a virtual environment. Bulletin of the Polish Academy of Sciences: Technical Sciences, 61(4): 871–884.

Wolbert, T. and Hegarty, M. (2000). What determines our navigation abilities? Trends in Cognitive Science, 14: 138–146. doi:10.1016/j.tics.2010.01.001.

Chapter **3**

User Modeling Systems Adapted to Virtual Environments

Graciela Lara López,[1,*] *Adriana Peña Pérez Negrón,*[1]
José Paladines[2] and *Francisco Rubio*[1]

1. Introduction

This work aims to highlight the importance of modeling a user within a virtual environment. The user model represents a set of personal characteristics related to a user or types of users. One of the objectives when modeling a user within a virtual environment is to study their behavior during the development of a real-world activity or task. To study human behavior in an activity simulated virtually implies considering elements like objects, actions or operations, in agreement with the specific conditions that are established within the environment.

The challenge of modeling a user is complex because effort is required to represent their skills, behaviors, and knowledge among other user's characteristics during the interaction with virtual environments. In this sense, the importance of user modeling lies in the fact of being able to create and modify a conceptual representation of the user, that is, to personalize and adapt the virtual environments according to the internal needs of the users. With this, it is intended to implicitly include the skills and declarative knowledge of each one of the types of users within the virtual system.

It is important to mention here that user modeling can be combined with techniques and methods of User-Centered Design (UCD), for the inclusion of more types of users in more contexts of use and not only virtual environments. User-Centered Design is a technique that is characterized by focusing on the processes of design and development. The main idea is to satisfy the needs of the user, and thus optimize usability of any system (Moreno et al., 2008).

A different context to virtual environments, for the purpose of training and/or instruction, are web projects. In any project, there is an analysis stage where the scope of

[1] Universidad de Guadalajara, Mexico.
[2] Universidad Estatal del Sur de Manabí, Ecuador.
* Corresponding author: graciela.lara@academicos.udg.mx

the website is defined. Likewise, from the information obtained, with some techniques of data collection, we gain the knowledge to start modeling users. With user modeling, we can identify and define the audience of websites.

In this context, the conceptual framework of the user modeling within virtual environments is presented; subsequently, we expose a set of projects of generic user modeling systems developed over the past twenty years and used in different contexts. Likewise, we describe their purposes and the different design requirements for the research prototypes.

2. User Modelling

Adapting a virtual environment to achieve the maximum degree of efficiency in the development of tasks or processes requires structuring of the contents of the environment in such a way that it meets the usability, accessibility, and user adaptability criteria.

In this sense, when tasks are performed within a virtual environment where the beliefs, interests, and characteristics of each user are organized into information sources represented according to a user model, it turns out to be a somewhat complex process.

To delve into the topic, a series of definitions of 'user model', given by different authors, are presented below:

Frias-Martinez et al. (2006) presented that the key element of a personalized environment is the user model. A user model is a data structure that represents the user's interests, goals and behaviors. A user model is created by means of a user modeling process in which unobservable information about a user is inferred from observable information about that user. Furthermore, they noted that the user model can be created by using a user-driven approach, in which the models are created directly from the information provided by the actual user, or an automatic approach, in which the user is shielded from the user model creation process.

For Lasierra et al. (2009), user modeling is the process of extracting and evaluating information that allows determining the behavior of a user.

According to Barla (2010), user modeling 'represents all kinds of information related to a user and the user's context, which are required in order to provide a personalized user experience. It can hold various features of the user, such as: age, education level, interests, preferences, knowledge, etc., or can represent the overall context of the user's work, including platform, bandwidth or location.'

Modeling computer system user's characteristics and attitudes have been a major challenge and a topic of special interest in the area of computing for several decades.

For Kobsa (2001), a user model is a set of information structures designed to represent one or more of the following data:

- Assumptions about one or more types of user characteristics in models of individual users (e.g., assumptions about their knowledge, misconceptions, goals, plans, preferences, tasks and abilities).

- Key common characteristics of users, pertaining to specific user subgroups of the application system (the so-called stereotypes) (Kobsa, 2001).

User modeling acquires information on the interests of users, their personality, their previous knowledge, their cognitive abilities, their beliefs, their learning preferences and progress. This is a means of generating and updating the user stereotype.

According to Fischer (2001), a user model is a representation of the user features with a view to decision making on computer system interaction.

On the other hand, Lozano et al. (2004) consider the need for representation of knowledge with the integration of intelligent characters, within virtual environments, to show the need for a representation of knowledge and the requirements that this entails.

Hassan et al. (2004) point out that user modeling serves as a support for decision making in the design of an environment, allowing the development of design centre more precisely on 'some' users. These users can be considered as 'real' because their description is based on a large group of real users.

Besides, Octavia et al. (2011) consider that, when interacting in a virtual environment, users are faced with a series of interaction techniques. These interaction techniques can complement each other, but in some circumstances, they can be used interchangeably. Due to this situation, it is difficult for the user to determine which interaction technique to use.

To adapt to the diversity of users, the construction of user models becomes crucial to provide adaptation and customization for users. User models contain information and assumptions about users and which play an important role in the adaptation process of user interfaces to the needs of different users. The user model is acquired through user modeling activity, which attempts to gather users' interaction patterns, preferences, interests and characteristics. The user modeling activity is acquired through a series of experiments that apply existing user modeling approaches through a user model template combined with individual user models. These authors proposed a conceptual framework as depicted in Fig. 1, to represent this information.

There are several levels of information provided in the user model. The general user model provides the most basic information that can be used for adaptation for all users. The group user model offers more specialized information to be applied to a group of users, while the individual user model provides more detailed information about a

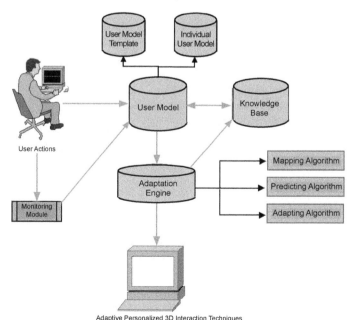

Adaptive Personalized 3D Interaction Techniques

Fig. 1: View of the conceptual framework for adaptive personalized 3D user interfaces by Octavia et al. (2011).

particular user. When the information is conflicting between the levels, more specific information takes precedence over more general information.

Next, we describe a number of works in chronological order, all of which seek to represent individual user characteristics in different contexts.

3. User Modeling Systems in the Last Twenty Years

The personalization of a virtual environment allows us to adapt the appearance and functionality of that environment according to the tasks, values, culture and needs of a user or a group of users. This customization allows the users to interact in a virtual environment previously organized and structured, where they can develop processes and activities, individually or together, in relation to age, values, social context, culture or other aspects.

In this context, we now describe a set of computational systems that integrate the user modeling in different research areas with some type of a virtual environment:

Santos Jr. et al. (2003) describe a dynamic user model to predict the objectives and intentions of an analyst during his massive information search tasks, providing an organization and presentation of the data. In this modeling, the analyst can proactively recover information that turns out to be novel and relevant, as required.

Their user modeling was part of a project called OmniSeer, which involved the company Global InfoTek, Inc. of the University of Connecticut and the University of South Carolina. The project focused on research, as well as technology evaluations and explorations of algorithms and necessary processes, to develop a framework of an architecture that will help analysts (users), not only to handle massive data, but also to collect prior and tacit knowledge. From an explicit approach, this user model allows the modification and updating of the analyst's current cognitive state.

In this dynamic user model the interests, preferences and cognitive context of a user are clearly outlined, as they change over time. This disintegration of the user model allows us to systematically construct and substantively analyze the behavior and performance of users of a project as sophisticated as OmniSeer. With this model, the interest and preferences of the analyst are stimulated, but it also allows explaining why an analyst wants to focus on certain given information.

The authors capture the interests, preferences and knowledge of an analyst to help in the tasks of searching for information, considering the beliefs of the users and the analyst's bias, see Fig. 2. The user model was integrated with a set of components which are described below:

1) *Networks of the User Model Component*: This component is composed of three basic subcomponents where interests, context and preferences are identified.

2) *Load of the User's Model*: During an analyst's information search session, the user model is initialized, considering a previous user model, to select a model template and thus represent a new analyst.

3) *Update of the User Model*: Based on the observation of the information search activities, the analyst's comments and other services of the user model are updated.

4) *Consultation of the User Model*: This component supports other services offering information on the analyst's current objectives and interests, providing signals of the intentions of an analyst.

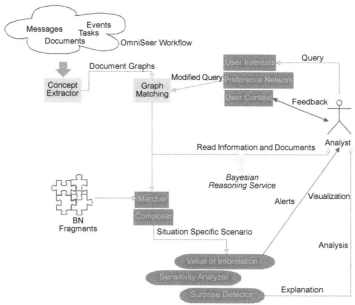

Fig. 2: Scheme user model in OmniSeer system of Santos Jr. et al. (2003).

5) *Explanation of the User Model*: It provides detailed comments to analysts and services in relation to the decisions taken from the user model, allowing the exchange of knowledge among analysts.

6) *Modification of the Analyst's Query*: Dynamically, the information queries of the analyst are modified and adjusted according to the user model.

7) Finally, the model is implemented with a Bayesian network engine, where the specific needs of each analyst and the interactions between them are adapted, with respect to their topics of interest and within the context in which the analyst's knowledge is developed.

Hassan et al. (2004) present an adaptation model based on techniques and methods of user centered design, for the insertion of different user types in more contexts of use. So it is possible to create a familiar environment ad hoc to the development of virtual environments, in specific designs on the web.

Their methodology is not reserved only for common interface solutions for users but is feasible for the development of multiple adaptable interfaces. Your adaptation model is described in the following three basic rules, where the steps of user modeling are considered:

1) *The design process should be focused on the user or all users*: The more information is known about the user, the better the design of the needs and characteristics of an environment can be adapted. Therefore, the environments may be more usable and accessible.

2) *The process of design, prototyping and evaluation are cyclical and iterative*: During the development of the environment, what is designed is evaluated through prototypes. Therefore, it will be less expensive to recreate a design than to completely redesign them.

3) *The user modeling and study process is done before any design*: The user modeling is reviewed and remodeled only when, in the evaluation process, design errors are discovered that may cause a defect in it.

In the user modeling, the user types of the environment are identified and defined. Following a methodology based on an inclusive design does not mean including all types of users, but users to whom the environment is not addressed should be excluded.

On the other hand, the authors of this model consider that the information of the users can be obtained through two sources: through surveys, interviews, observation, etc., or through studies and literary analysis on user conditions and needs.

The user centered design, as well as the user modeling, must be directed by the needs and objectives of the user or of all the users, so that the information obtained should be useful and manageable for the designer of the environment. For that, it is necessary to schematize and organize the information to perform user modeling. With this, user profiles are defined, with the aim of covering the needs of each group of users. Profiles are defined on the basis of common attributes and limitations among users, to be grouped later.

In this sense, there is no doubt that in recent years, more information is available on the web and interest in the personalization of users has grown. Therefore, the development of intelligent agents has provided assistance based on the contents of the web and the interests of the users. Intelligent agents take into account the knowledge of users that is obtained from their profiles.

An agent is designed according to the preferences and interests of the user, through observation of the behavior of the users themselves. With this, it is possible to acquire and model categories of interest of users.

On the other hand, the authors consider that a problem for the user modeling occurs when the people are too many, because they have different behaviors. In these cases, it is convenient to make use of user archetypes, that is, an exemplary user from which other users are derived. Thus user modeling based on the definition of archetypes represents patterns of behavior, objectives and needs that can be used for decision making in the website. This could help the designer to make a design focused on the user, or on more than one user.

It is common to find that the designer cannot understand why someone finds complicated the use of their website; these archetypes of users make the designer to have in mind a 'real' user, with limitations, skills and real needs.

On the other hand, Gascueña et al. (2005) presented a student-centered model, using two ontologies that work simultaneously—one ontology stores learning material, and the other contains the student profile (Fig. 3). The main purpose of this model is to reuse learning materials and adapt teaching methods based on the user. An agent operates on both ontologies, selecting and showing students the items of the educational material that better match their profile.

The student ontology organizes data around a class called 'student'. This class contains student data, like visual preferences, devices, personal data, courses in which the student is enrolled and connection speed. The idea of this organization is to adapt learning as close as possible to the student.

Godoy et al. (2006) proposed a clustering algorithm, called WebDCC, which allows the creation of user profiles in intelligent information agents. The algorithm develops an incremental approach, making it easier for agents to interact with users over time,

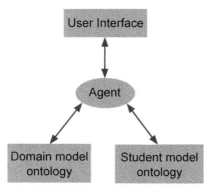

Fig. 3: Adaptive architecture to the user by Gascueña et al. (2005).

to obtain and adapt the user's interest categories. Agents can treat unpredictable subject areas. In addition, this algorithm manages to focus on aspects of categories in which the user is really interested.

The forms of user profiles in this algorithm provide clear grouping options that can be simply interpreted and explored by users and other agents. By extracting the semantics of web pages, the algorithm generates intermediate results that can be integrated into a format that is understandable to the machine, as in ontology. Figure 4 illustrates a user behavior, with a personal search that possesses a number of reading experiences in the Visual Basic and Java programming languages, along with some readings about soccer and motorsports.

De Rosis et al. (2006) developed a system that dynamically models an agent and users, with the objective of having a conversation through a text dialogue without restrictions, so that users change their eating behavior.

The authors highlight in their work the importance of conversational systems, which are highly supportive in areas of medicine, especially in counseling about eating behavior. Counseling is provided by an embodied conversational agent (ECA), who needs to be in tune with the emotional state of the patients. In this sense, adaptability comes

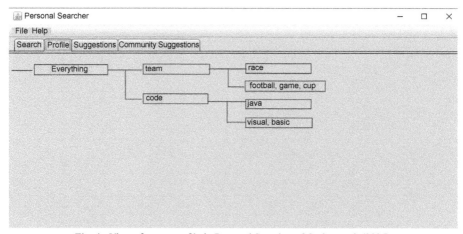

Fig. 4: View of a user profile in Personal Searcher of Godoy et al. (2006).

to be a particularly relevant characteristic for its implementation, from a perspective of accommodation to the counseling dialogue. For this reason, its conversion system considers the following characteristics:

- A dynamic user modeling, in which the user's state is identified and this image is examined during the interaction.
- A dynamic agent modeling is where the emotional reaction of the agent is personalized before the response of the users.
- A double adaptation of the dialogue is made to the user and to the agent. Within the dialogue system, the expressions are given by an ECA.

With these elements and after having made a meticulous analysis of several dialogues that occur naturally, a dialogue simulation system is proposed to challenge the behavior of humans when they communicate. Figure 5 presents the structure of the dynamic user model with the characteristics described below.

Peña (2007) presented a student model based on cognitive maps in his doctoral thesis. The system analyzes students in order to ascertain their interests, skills and attributes, and thus create a mental map. The mental map (also called cognitive map) has the function of representing the student's ideas and their interrelationships.

The user model (in this case of the student) takes into account their profile, preferences, skills and their evolution over time. Modeling is a two-part process:

1) The concepts that integrate the mental image are represented in a cognitive map.

2) These concepts are defined in an ontology.

The ontology represents the student's knowledge domains. Seven domains were defined: three for student characterization (cognition, personality and preferences), three more for experience characterization (sequence, content and evaluation), and another one for the student knowledge acquisition management.

Another case based on agents is the one proposed by Lasierra et al. (2009). They developed an SNMP agent v3 (Simple Network Management Protocol) for the modeling of users in LAN (Local Area Network) environments, with the purpose of modeling users, according to the information contained in the MIB (Management Information Base). The

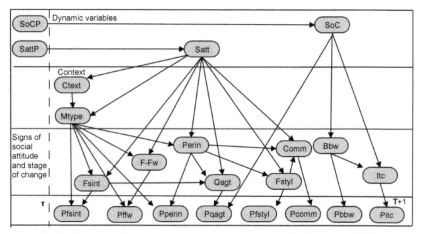

Fig. 5: Structure of the dynamic user model proposed by De Rosis et al. (2006).

general purpose of this development is to locate a pattern that characterizes the behavior of users.

The agent development offers a tool to model all the users that connect to the LAN network under the SNMP architecture. For their model, they considered a process that consists of three phases: (1) selection of information, (2) characterization and learning of behavior, and (3) supervision and detection of anomalies. These phases are described below:

1) *Selection of information*: The data considered by the authors, to learn the behavior of the users in which the objective is to classify the users in different categories or profiles was based on two parameters: On the one hand, the distribution of resources in the internet access, such as the width of band used, the duration of the connections, the size of the packets sent and the ports used to identify the use of peer-to-peer applications. On the other hand, aspects of traffic, such as the number of packets sent and received, erroneous packets and analysis of headers or numbers of connections, among others; in addition to two internal parameters, such as the use of memory and CPU.

2) *Characterization and learning of behavior*: Through an SNMP management architecture, the user's information is collected for subsequent exchange. This is a flexible and feasible solution, frequently used by user modeling systems. The information collected from each user is processed and classified as normal or usual for the modeled user. During this phase, it is considered a learning time, where one can observe and define, in a representative way, the different patterns that each user can present.

3) *Supervision and detection of anomalies*: In the testing or supervision phase, survey periods are established to obtain user information, compare their behavior and classify it as normal. In this way, they seek to detect anomalies in the monitored behavior.

Another interesting model, presented by Conati and Maclaren (2009) is a model for recognizing user emotions during participation in an educational computer game. The model considers a high level of uncertainty in the recognition of a wide variety of emotions of users, between the causes and effects of emotional reactions.

Considering the emotions that a user feels and the motives, the authors based their model on the OCC cognitive theory of emotions—a theory that shows that emotional knowledge is organized hierarchically with a first factor that divides emotional states into pleasant and unpleasant (Ortony and Terence, 1990). Under the design of an agent that responds to emotions, it can recognize when a user feels a negative emotion, such as shame for having done something wrong. At the same time, the agent offers the user with some suggestions to make him feel better about himself. In another case, the agent recognizes when a user is upset by his/her behavior, due to some reproach. In this case, the agent provides support to amend it.

Considering that the OCC theory captures instant emotions to specific events, as opposed to affective states of frustration, boredom or confusion, the authors evaluated the addition of diagnostic information, through an affective user model, to detect the emotions of players. Therefore, their model combines information on the cause and its effects of users to recognize multiple emotions that change quickly. The importance of the model lies in the combination of a diagnosis and predictive inference. Figure 6 shows a

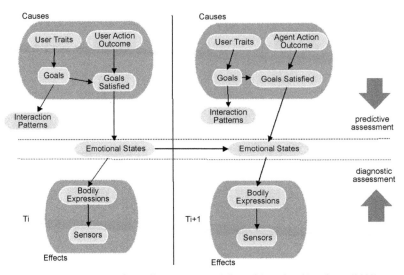

Fig. 6: Representation of the affective user modeling of Conati and Maclaren (2009).

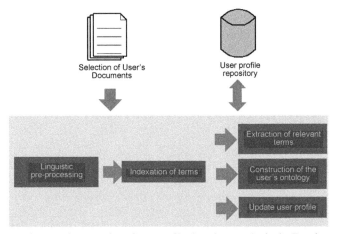

Fig. 7: Process of automatic generation of user profiles based on ontologies by Ferreira et al. (2010).

network that can combine evidence on both the causes and effects of emotional reactions to assess the user's emotional state after each event.

Ferreira et al. (2010) represented user profiles, using ontologies with fuzzy logic. Users of a learning platform can publish, create and collect specified content. The system constructs an ontology following a five-phase process, as shown in Fig. 7.

1) The *linguistic preprocessing phase* is responsible for collecting and converting the different files uploaded by the user into a common internal format for proper processing in the subsequent steps.

2) The *term indexation phase* adds a small index used later to build the ontology. This index stores an identifier associated with each term and a list of references to the respective term for each entry.

3) The potentially most interesting terms are selected from the above collection in the key *term extraction phase*.

4) The *user ontology* as such is constructed in the following phase. The ontology is considered as a set of related trees where each node represents a topic. The idea is that the resulting tree is a taxonomic representation of previously filtered concepts.

5) The *user profile is updated in the last phase*. Here new data is added to the ontology as contents published by each user. To do this, the added documents are processed as above, and the ontology is later modified, based on the resulting information.

On the other hand, through techniques based on initiatives, Emond et al. (2010) presented the specifications of requirements for a virtual training environment of users, through a user modeling methodology, based on three concurrent processes: task analysis, the specification of simulated firearms and cognitive modeling.

In the task analysis, the performance objectives for training, related to training techniques based on Close Quarter Battle (CQB), are specified. These Close Quarter Battle (CQB) techniques consider the existence of echelons of situational awareness within a space. The individual must have a deep understanding of things that are close to themselves and those that occur in other areas of the nearby room or building. In this sense, the authors also declare that it is possible to perform different types of task analysis, including hierarchical analysis (also known as prerequisites to tasks), and procedural analysis (also known as information processes or cognitive task analysis).

In the case of task analysis, a cognitive modeling activity is carried out for the development of constructive simulations of CQB skills. This study is done to support the virtual training environment, the Immersive Reflexive Engagement Trainer (IRET), which is a collaborative research effort between the Canadian Department of National Defence (DND) and the National Research Council of Canada, Institute for Information Technology (NRC-IIT). This trainer allows combining of a number of existing technologies for soldiers to train simultaneously within virtual and real environments.

Related to the cognitive modeling of the IRET project, two objectives were established: (1) to develop high fidelity cognitive modeling technology to integrate as artificial intelligent agents in an immersive combat game; and (2) to develop detailed models of student's performance and learning to support the instructions.

To achieve the first objective, the development of an agent with perceptive, motor and communication abilities was thought of. In addition, this agent is autonomous and object-oriented, and is placed in an environment in which it interacts with other agents. He is also flexible and has the ability to learn and adapt his behavior based on his experience. For the achievement of the second objective, the simulation of firearms and the related user interface technology were considered to provide trainees with realistic input devices for the training simulator.

Figure 8 shows a flow diagram of the cognitive modeling methodology, spanning from task analysis to model verification and validation. Processes are represented as ellipses and products as rectangles.

Schoefegger et al. (2010) proposed a user model based on the context, in relation to the emerging topics faced by the participants, and their previous levels of knowledge within a learning system. The authors took into account the experimental data from users who were learning a topic with the help of a collaborative labeling system. Likewise, authors relied on an intelligent automated support, which captured the challenges in integrated learning in the work of various dimensions.

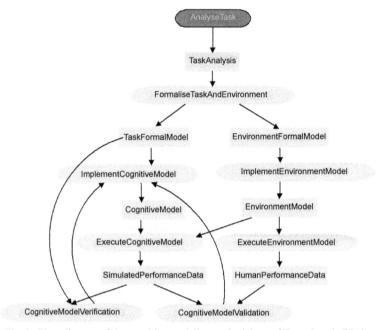

Fig. 8: Flow diagram of the cognitive modeling methodology of Emond et al. (2010).

From the above, personalized recommendations are made, according to the context of each user. These recommendations are fundamental supports to help knowledge workers during their work as researchers or students, achieving, this way, to model each user according to their interests and abilities, within a WIL system (Work-Integrated Learning), with respect to the management of their own resources, and the search for material that may be relevant, depending on their tasks.

In the development of this user model, the authors considered the interpretation of the data on the use of the system, at the same time used heuristics to organize, share, diagnose and increase the levels of knowledge of the users, in each of the topics available with a static knowledge domain.

For the development of the profile of users in dynamic and adaptive WIL systems, the authors chose to use a collaborative tagging system, because this is a system that shows highly dynamic user behavior and emerging topics of interest. These collaborative labeling systems have become popular support tools within user modeling since they allow observing the combination of their activities in the learning systems over time and knowing how they are made.

For the identified topics and mentioned entities in tweets, Abel et al. (2011) proposed a user modeling framework based on the activities of users and which enrich the semantics of Twitter messages. The authors analyzed strategies for construction hashtag-based. Furthermore, they measured and compared the performance of the user modeling strategies in the context of a personalized news recommendation system. The results revealed the variety and quality of the generated user profiles and how user modeling strategies impacted.

With more 190 million users and more than 65 million messages of Twitter per day, Twitter is the most prominent micro-blogging service available on the web. The user modeling strategies considered a design of three dimensions: (1) the type of profiles created by the strategies, (2) the data sources exploited to further enrich the Twitter-based profiles, and (3) temporal constraints that are considered when constructing the profiles.

Furthermore, a dataset of more than 2 million Twitter messages user's profiles was created and revealed several advantages of semantic entity and topic-based user modeling strategies, among which stand out enrichment with the semantics extracted from the news articles, which correlate with the activities of users within Twitter. Furthermore, varieties of built profiles are developed and enriched the accuracy of the news article recommendations, significant for users.

On the other hand, a framework proposed by Kaklanis et al. (2011) is based on a new virtual user modeling technique which describes in detail all the physical parameters of a user with disability(ies) to generate a dynamic and parameterizable virtual user model, which is used by a simulation framework to assess the accessibility of virtual prototypes.

Showing how the proposed framework can be used in practice, the authors present an evaluation scenario where a car designer simulates accessibility and try a prototype for different virtual users with disabilities. The designer initially develops a virtual workspace, using a virtual user model editor. For this model all the possible disabilities of the user are considered, such as motor, visual, hearing, speech, cognitive and behavioral; user characteristics; as well as some general preferences (e.g., preferred input/output modality, unsuitable input/output modality, etc.).

The designer generates two cases of virtual user model with physical different (a user with advanced age and a user with spinal cord injury) characteristics. Two task models, that describe the handbrake use and the opening of the storage compartment, were developed and tested respectively. To try the interaction of the virtual user in the specific virtual environment, a simple simulation model was described, which allowed demonstrating the problems of accessibility to the interior of a car.

A similar case to the previous one is presented by Moschonas et al. (2011). They evaluated several ergonomic factors on a framework that performs automatic, simulated ergonomic analysis of virtual environments. The framework considers a virtual user model describing users with or without physical deficiencies and evaluates the ergonomy of virtual environments for specific users. Two new ergonomic factors related to comfort are presented and compared to meet physical metrics, such as torque, momentum and energy.

The human factors were applied into two different realistic scenarios—one coming from the automotive area, and the other considering a workplace. In each scenario, two different designs were tested and their ergonomics were compared. For every user model, the two design prototypes of each scenario were compared according to the proposed human factors. The proposed factors were recorded at six body locations: lower torso, middle torso, upper torso, shoulder, elbow and wrist, the reference arm was the right one for the automotive and the left, for the workplace scenario.

Skillen et al. (2012) developed an extensible user profile model through the creation of a user profile ontology for the personalization of context-sensitive applications in mobile environments. They considered analyzing users' behavior, characteristics and needs. In addition, they placed special emphasis on the ontological modeling of dynamic components for their use in adaptive applications. They found that the modeling of the user profile is essential for any type of customization, retrieval of customized information, customizable user interfaces or customized delivery of context-sensitive application services.

Considering that a user profile is a digital representation of the unique data relating to a particular user, the modeling of the user profile involves the creation of a data structure that can contain these characteristics/attributes of the type of user. This data structure is generally a template to generate specific user profiles for different individuals.

The authors considered that ontologies represent a controlled vocabulary which is structured into a hierarchical taxonomy, where the key domain concepts are found. Each defined class of ontology may have parent-and-child classes forming a hierarchy of related concepts. Properties exist in each class, describing features of the class and any restrictions placed upon them.

A distinctive feature of this model is its focus on dynamic and temporal concepts to allow customization of applications as the user moves between mobile environments, where additional evaluation will be subject to the model being fully implemented and deployed within smart, adaptive applications. Figure 9 illustrates the ontological structure with a hierarchy of user ontologies.

Clemente et al. (2014) presented the design of mechanisms of the student model to deal with the non-monotonic nature of the pedagogic diagnosis. This proposal was validated in the context of a virtual laboratory of biotechnology where one or more students can perform a practice of genetic engineering.

Through a student modeling agent, which is part of a multi-agent system known as MAEVIF (Model for the Application of Intelligent Virtual Environments to Education) of the Polytechnic University of Madrid, information is obtained about the student and it provides a tutoring agent, who supervises and guides students during their learning process, with the support of a virtual environment.

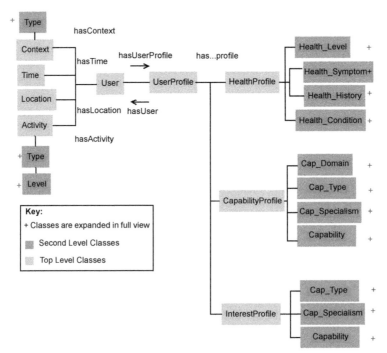

Fig. 9: View of user profile ontology that shows the ontology hierarchical structure of Skillen et al. (2012).

Within the virtual environment, the student can move from one place to another and can perform actions, such as lifting objects or pressing buttons. These actions are scheduled with a set of preconditions, which are sent to the tutoring agent and the student agent, to keep track of the student's activities.

It is important to mention that this design was supported by two data ontologies: (a) student ontology, and (b) world ontology. The first contains information on the structure and content of the virtual environment; the second represents the student's information, like profile, behavior within the virtual environment, beliefs and the state of knowledge.

Quiroz et al. (2015) considered that the use of information technology and communications (ICT), especially with the use of Learning Management Systems (LMS), for allowing teaching expansion beyond the boundaries of the classroom. For this reason, LMS searches are used for designing innovative virtual learning environments (VLE), placing the student at the center of the educational process. For this, a methodological change is necessary to move from a traditional approach, focusing on the content and the teacher, to one focused on e-activities and the student use of active methodologies.

For this reason, the model proposed by the authors is based on methodological learning focused on the student, where conviction between the people is established that learning takes place by doing and interacting. The development of activities that help teamwork and generate a recreation of real-life situations is encouraged. The model is centered on the student to find, in the e-activities, the specific indications that must be carried out, being the basis of the rest of the resources as presentations, documents and tools of the platform that are necessary to facilitate learning. The structure of an e-activity is presented as a website and displays description, objectives, time of development, indications, how to evaluate and references.

Figure 10 shows how a set of elements are taken into account to design three-dimensional learning objectives, which can be useful to strengthen an effective learning process, with a practical and attitudinal axis of the students. Linked to this, the authors seek an active process of learning based on concepts, skills and attitudes from a three-dimensional interface, that is, the learning process is directed towards practice.

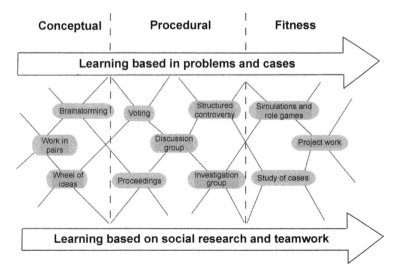

Fig. 10: Three-dimensionality of learning objectives of Quiroz et al. (2015).

Likewise, authors consider that designing and implementing of virtual learning environments (VLE), focused on the students, remains challenging due to the fact that time and effort are required to consider the particular characteristics of each student. However, the VLE is a tool that allows interaction, communication and multimedia, which the student develops outside the classroom.

Under another line of research, Aballay et al. (2015) studied how people from diverse countries have dissimilar cultures and interact with the user interface in a different way. The users prefer graphic designs and have their own expectations and behavior patterns. Therefore, the user interface should be adapted to the needs of each geographical location to facilitate an optimal user experience.

The decisions of adaptation in adaptive web systems are based on the user's characteristics represented in user modeling. The compilation of data of students to determine their characteristics is an important factor for the development of an adaptive learning system. Another point is the culture-model, where the culture represents beliefs, values and symbolic significances that the people have in a group or community. There are different culture-models with different patterns and guidelines, out of which the authors proposed seven to web design.

The first is the translation which is the basic process of internalization applied to web design; the second is the color which has different meanings in different countries; the third is the direction of text and symmetry respecting the language characteristics; the fourth is collectivism individualism on how people work; the fifth is feminism and masculinity; the sixth is contextualization of the differences between cultures; seventh is architecture of information and how the information is distributed on websites.

From the perspective of learning, Uribe-Rios et al. (2016) found that students with high capacities can have innate creativity for learning faster, a higher IQ, among other characteristics. But, why do some have problems with their low academic performances? It is believed that the solution to the problem is motivation. This motivation can be enhanced by involving the student in the creation of educative material. The idea is proposed as the design of a profile's model that supports an adaptive process of co-creation of educational material based on the characteristics of students with high capacities.

In this sense, to understand and work out a process of co-creation of educational material, a profile model centered on three aspects is proposed: individuals involved in the co-creation, teaching and learning models, and the adaptation to the process of co-creation. The family that influences the student into an adaptation process, especially as a personal profile, offering new forms of work for education adaptation and the definition of profiles for the co-creation and high capacities, is considered. This is not an easy process due to the little work developed in the field and the differences in each student boosted by high abilities; so this is considered as an atypical adaptation process.

Taking into account the several aspects for the co-creation of educational material, such as adaptation, teaching-learning models and the individuals involved, this allows correlating the high capacities and the said material directly, since it frames this from the definition of the profiles of adaptation. In this way, the design of the profile model enhances the individual characteristics of the students, encouraging their motivation towards learning.

Figure 11 presents the profile model proposed by Uribe-Rios et al. (2016), where the profiles of the student, the family, the teacher and the context of co-creation are observed; so is the social profile that depends on the teaching-learning models and the characteristics of each student with high abilities.

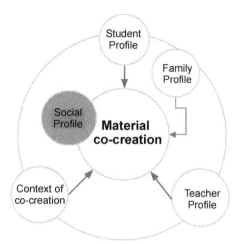

Fig. 11: Profiles model by Uribe-Rios et al. (2016).

The main technological support for adapting learning contents is the management of students' characteristics through the user modeling, according to (Silva, 2016). He felt that the adaptation is essential in whatever learning virtual environment it is, given that it is an evolutionary process that depends on the characteristics of the apprentices and their evolution. With these characteristics, the users demand an adaptive environment to their specific needs and preferences. Further, the Instructional Management System Learning Design (IMS-LD) that he proposed, focuses on the transformation of educational contents to the unit's interoperable learning in any virtual environment.

With this approach, users must be modeled according to their preferences in technologies and electronics resources. For example, when the students enter a platform, the system displays a questionnaire that the students must answer. All that information will be saved as properties for the user profile and this facilitates the content to be adapted according to their values. With this information, the system can offer the students an adapted learning path to their specific needs. This method allows the learning management system (LMS) to create, read and update objectives, interests, competitions and user preferences.

Lara (2016) presented, in her doctoral thesis, a computational model for the generation of directions for the location of the objects in virtual environments. In this thesis, she tries to be as adaptive as possible to the model of the user and to the environment. Her user model seeks to represent the user in a computational manner with an ontology called 'user ontology'.

Using the scheme proposed by González (2014), the 'user ontology' was developed:

1) Identification of the relevant characteristics of the user.
2) Creation of the conceptual model of the user ontology.
3) Implementation of the user ontology with the Protégé tool.

She analyzed and selected different types of characteristics:

1) *Basic characteristics*: Essential attributes that identify each user. These attributes can also be called personal data and contain static information about the user, such as name, age, gender, email and previous training, among others. This set of

characteristics is used for administrative purposes, except for the previous training data that will allow adapting the system to the previous knowledge of the user in specific domains, such as chemistry, computing, etc.

2) *Cognitive characteristics*: The general capability of remembering the spatial location of objects was included, as it is useful to determine the type of directions that are more adequate for the user.

3) *Perceptual characteristics*: Including characteristics that allow the system to infer the way in which the user visually perceives a scene, such as visual acuity and blindness of the color.

4) *Knowledge and experience characteristics*: Representing the knowledge the user has of the specific objects and object types in the environment, as well as the area or areas of knowledge of the user.

The conceptual model of the user ontology was created (Fig. 12).

Figure 13 presents a conceptual map of the user ontology.

The principal class of user ontology is the 'User' and it is related to three classes: UserCharacteristic, KnowledgeOfField, and Profile.

- The class 'profile' contains all the personal information or basic characteristics of the user.

- The class 'UserCharacteristic' has two sub-classes that represent: (1) the cognitive characteristics and (2) the perceptual characteristics.

- Finally within the subclass 'CognitiveCharacteristics' there is a subclass called MemoryOfLocation. The subclass 'PerceptualCharacteristics' has two subclasses called: ColorBlindness, and VisualAcuity.

Within the user modeling, a topic of special interest is the development of models oriented towards the recognition of emotions in virtual environments.

According to Guojiang and Liu (2017), emotional intelligence is an important component to improve the credibility of a virtual character's behavior, given that expressing appropriate emotions can strike a communication between virtual character

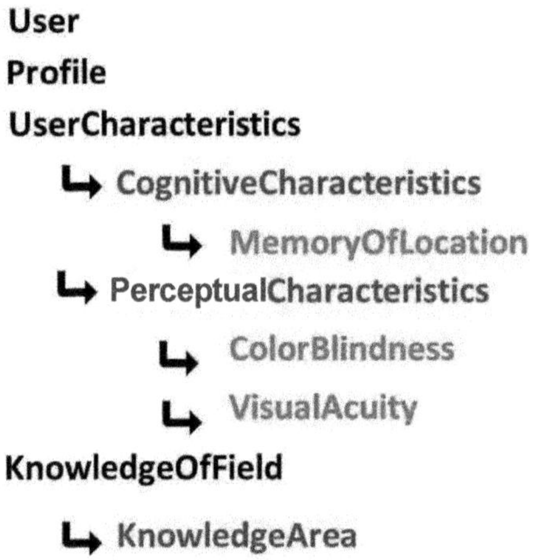

User
Profile
UserCharacteristics
    ↳ CognitiveCharacteristics
        ↳ MemoryOfLocation
    ↳ PerceptualCharacteristics
        ↳ ColorBlindness
        ↳ VisualAcuity
KnowledgeOfField
    ↳ KnowledgeArea

Fig. 12: Hierarchy of classes of the user ontology of Lara (2016).

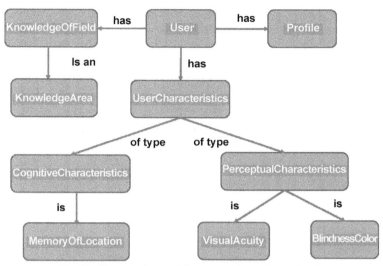

Fig. 13: Conceptual map of the user ontology of Lara (2016).

and a human more comfortable, based on psychology and artificial psychology; achieving this will simulate interaction between human emotions, needs and personalities. The results of the proposed method by the authors show that it can make the virtual character to produce a more natural emotional reaction during the interaction.

With 10 emotions, according to their intensity, 5 positive emotions, and 5 negative emotions, they evaluate the emotion of a virtual character, considering that the personality is the dynamic organization of the physical and mental system that determines the behavior and the unique thought of the individual. The emotional virtual character system proposed by them includes the perception module, the emotion module and the behavior module. The perception module includes a sensor of information mainly composed of recognition of facial expression and transforms the recognition results into an emotional measurement to transmit to the emotional system. The emotion module has three subsystems: need, emotion and personality; they determine the type and intensity of the emotional reaction. The behavior module relies on the information passed by the emotion module to select an appropriate behavior.

Taking up the case of the consultation of documents by users Kotzias et al. (2018), through an investigation, it was discovered that individuals usually consume a few items from a large selection of items. According to the general characteristics of article consumption, it is observed that this consumption is repeated and novel. In this context, the traditional methods of matrix factorization frequently used have some disadvantages in reconstructing the individual characteristics, which cause a lack of predictive precision at the user level. With this analysis, the authors propose a method based on mixed models, with a simple and scalable approach for user modeling in this context.

The authors use the Expectation-Maximization (EM) algorithm to learn the global and individual mixing weights. Validation of the model is done in terms of accuracy in user consumption predictions by using several real-world datasets, including location data, social media data, and music listening data. Experimental results show that the mix-model approach is systematically more accurate and more efficient for these problems as compared to a variety of state-of-the-art matrix factorization methods.

A useful aspect of the mixed model approach is that additional components can be included to potentially improve its predictive power, instead of relying on a single population component to generalize a user's model toward global population patterns.

An interesting case is also the one presented recently by Wang et al. (2018, July). They developed an Astronaut Modeling and Simulation System (AMSS), in which the performance in physical and cognitive tasks during space flights is analyzed. AMSS is the first tool in China, where human systems for space missions are modeled and designed. In this system, several factors that cause stress in astronauts were considered, such as microgravity, confinement and radiation, which can affect human abilities. The main goals of this project were to gain a better understanding of the capabilities of the astronauts and to better predict the human performance of tasks during space flight.

There are three important aspects of the astronaut's performance modeling: (1) how to use human biomechanical and cognitive parameters to characterize the individual differences, and how these parameters change in spaceflight environment; (2) how to describe the task and the interaction between the human-system, such as the type of task, the procedure of the task, the human-machine interfaces, and demands of the task; and (3) how to evaluate the effectiveness of the human system in a precise and quantitative way. Figure 14 presents a three-level model architecture: human characteristic model, behavioral model, and performance evaluation model.

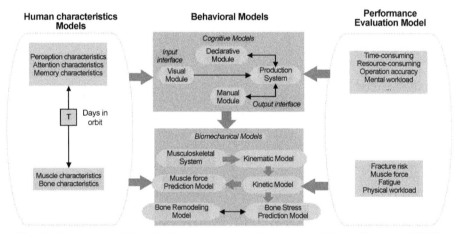

Fig. 14: Three-level model architecture for astronaut performance modeling by Wang et al. (2018).

4. Techniques to Support User Modeling

For the personalization of user characteristics in the system, we detect there are several techniques, tools and processes for user modeling. Among the most common techniques that have been used for the representation of user modeling are the following: direct observation, interviews, questionnaires, Bayesian, based on stereotypes or inference rules. Next, each of these techniques is described in a general way.

1) *Direct observation*: Oriented to the characterization of user classes and linked tasks, this technique considers critical aspects, such as social aspects, that can be incurred in the behavior of users, when the environment projects a specific context.

2) *Interviews*: Through the collection of personal experiences, opinions or motivations of behaviors. These are of great value to identify previous knowledge and difficulties in the frequent use of tools or development of particular tasks.

3) *Questionnaire*: Through a set of open or closed questions during a directed or non-directed interview, it is possible to obtain a general perspective of a situation with limited answers.

4) *Bayesian*: Based on probabilistic studies, a hypothesis is tested to prove a certain event.

5) *Based on stereotypes*: Through a simple method of acquisition, users are classified into categories, to make predictions based on them. All users belonging to the same category are considered similar in their characteristics and/or behaviors, under a certain set of circumstances in the same context.

On the other hand, Viviani et al. (2010) presented two approaches. In the first approach, the use of standard ontologies or unified user models is considered, that is, user modeling based on standardization. In the second approach, there are the mappings between different representations of user models, which are characterized as a model based on mediation.

The ways of gathering information about a user and their context in a user model, and at the same time, the possibility of sharing information between different systems centered on the user, have allowed the classification of the two approaches mentioned above and which are described below:

User modeling based on standardization considers a top-down view, where some *a priori* standards focus on the applications involved that must be met.

The user model, based on mediation, is distinguished by an ascending behavior that operates with the representation of different representations of user models, where the data of the user model is transferred from one representation to another in the same domain or between domains.

Table 1 singles out six important aspects of the selected documents for the documentation of this chapter. Column 1 lists the year of publication, Column 2 gives the authors of analyzed papers, Column 3 gives the name of project where the user model is presented, Column 4 describes the purpose to which the user model was oriented, Column 5 shows the modeled features of the user and Column 6 specifies the technique used by the authors for their user modeling.

As a final conclusion from the analysis of the state of the art in modelling user, presented in Table 1, we can observe that all the systems used a virtual environment, albeit with different contexts (i.e., learning platforms, websites, or others).

The main goal of nine of the analyzed systems is student modeling for the use of virtual learning environments. However, all systems were analyzed in search of key aspects for user modeling, regardless of the purpose for which they were modeled. Four of the 20 systems developed are oriented towards information seeking of users in websites. Three systems promote health in terms of physical disabilities and feeding.

The reviewed papers described different modeled user characteristics, such as interests, abilities, preferences, habits, needs, disabilities or individual limitations, learning styles, positive and negative emotions, behavior, physical aspects, competences, among others.

Table 1: User modeling systems adapted to virtual environments.

Year	Author	Name of Project	Oriented to	Modeled Features	Technique Used
2003	Eugene Santos Jr., Hien Nguyen, Qunhua Zhao and Hua Wang	OmniSeer	Information seeking of user	Interests, preferences and cognitive context	Bayesian network engine
2004	Yusef Hassan Montero and Francisco Jesús Martín Fernández	Unknown	Development of usable and accessible websites	Needs, objectives and characteristics with some type of disability or individual limitation	Unknown
2005	José M. Gascueña, Antonio Fernández-Caballero, Pascual González	Unknown	Adapt the teaching to the individual characteristics of each student	Student's learning style (active, reflective, sensory, intuitive, visual, verbal, sequential, global)	Data ontology
2006	Daniela Godoy and Analía Amandi	WebDCC: the Personal Searcher agent	Information seeking task on web	User interests	Clustering algorithm with intelligent information agents
2006	Fiorella de Rosis, Nicole Novielli, Valeria Carofiglio, Addolorata Cavalluzzi, Berardina De Carolis	WOZ	Promote health	The emotional state of the user	Bayesian network with embodied conversational agents
2007	Alejandro Peña Ayala	Student model in the Web-based education systems	Know the causal effects that a teaching-learning experience will produce on the student	Interests, skills and attributes	Cognitive maps and ontologies
2009	N. Lasierra, M. Lopez, J. Garcáa, A. Alesanco	Unknown	Have control and supervision of behaviors of users in LAN environments	Habitual behavior of a user in a network	For an agent
2009	Cristina Conati, Heather Maclaren	Unknown	Recognize multiple emotions of users	Emotions, causes and effects	Using an EMG sensor
2010	Mateus Ferreira-Satler, Victor H. Menéndez, Francisco P. Romero, Alfredo Zapata, Manuel E. Prieto	AGORA	Management of learning objects within a LCMS (Learning Content Management System)	User preferences and organization of information	Blurry ontologies

Year	Authors	Description	Objective	User characteristics	System type
2010	K. Schoefegger, P. Seitlinger, T. Ley	This user model will be performed within the MATURE project	Make personalized recommendations in integrated learning work	User's topics of interest and knowledge level	Collaborative tagging systems
2011	Fabian Abel, Qi Gao, Geert-Jan Houben, and Ke Tao	Unknown	Analyzing User Modeling on Twitter for Personalized News Recommendations	User's profiles and entities people refer to (persons, products, etc.)	Hashtag-based and entity-based profiles
2011	Nikolaos Kaklanis, Panagiotis Moschonas, Konstantinos Moustakas, and Dimitrios Tzovaras	Simulation module, co-funded by the project VERITAS	Users with disabilities	The affected by the disabilities tasks, motor, visual, hearing, speech, cognitive and behavioral user characteristics as well as some general preferences (e.g., preferred input/output modality, unsuitable input/output modality, etc.) of the user	Framework with user virtual in a virtual environment
2011	Panagiotis Moschonas, Nikolaos Kaklanis, Dimitrios Tzovaras	Framework that performs automatic, simulated ergonomic analysis of virtual environments, co-funded by the project VERITAS	Users with or without physical deficiencies	Six body locations: lower torso, upper torso, shoulder, elbow and the wrist	Framework with user virtual in a virtual environment
2011	Yusef Hassan, Francisco J. Martin Fernández, and GhzalaLazza	Unknown	Web design centered in the user	Patterns of behavior, objectives and needs	Website
2012	K.L. Skillen, L. Chen. C.D. Nugent, M.P. Donnelly, W. Burns, I. Sholheim	Application personalization within mobile environments	User profile Modeling	Users' behavior and characteristics in context-aware applications (Habits, preferences and interests)	User Profile Ontology
2014	Julia Clemente, Jaime Ramírez and Angélica de Antonio	Design of mechanisms of the student model to the pedagogic diagnosis for project MAEVIF	Student profile Modeling	Behavior, beliefs and the state of knowledge of Student's	User Profile Ontology with an Intelligent Agent

Table 1 contd. ...

... Table 1 contd.

Year	Author	Name of Project	Oriented to	Modeled Features	Technique Used
2015	Juan Silva Quiroz, Elio Fernández Serrano, Andrea Astudillo Cavieres	Learning Management Systems (Module)	Design of virtual learning environments	Name of the activity, purpose; motivation, number of participants, time for the participant, time for the tutor and evaluation	Website
2015	L.N. Aballay, S.V. Aciar, C.A. Collazos, and C.S. Gonzales	Proposal for the project 'Development of Technological Tools to Support Virtual Education'	Adaptation of ubiquitous environments for collaborative learning according to the cultural profile of the user/student	Language's characteristics, color, contextualization, architecture of the information and sex	Website
2016	Maria del Mar Saneiro Silva	Psychoeducational Bases in the Design of Courses Implemented in IMS-LD for Virtual Learning of Students with Functional Diversity	Adaptation and personalization of contents, resources and learning flows	Demographic data: Personal information: (name, date and place of birth...), Psychoeducational preferences: (attention deficit, language skills, time management, etc.), Preferences: (language), Learning styles: (perception, processing, comprehension, sensory input), competencies: (level of collaboration, communication, participation, initiative, etc.), progress: (level of knowledge acquired, objectives, interests for each learning object, etc.)	Central storage repository

Year	Authors	System	Description	
2016	Mery Yolima Uribe-Rios, Juan Pablo Meneses-Ortegón, Teodor Jové, Ramon Frabegat	Unknown	Adaptation in the process of co-creation of material for students with high capacities and some teaching-learning models	Context profile of co-creation: contains those characteristics of the environment, the tools, devices and resources that intervene; Profile of the teacher: it contains basic characteristics and the teaching of the teacher. Family profile: contains the most outstanding family aspects that directly affect co-creation. Student profile considers cognitive, learning, socio-emotional and risk variables — Personal profiles
2017	Wang Guojiang, LuiJie	Unknown	Emotional modeling for a vision-based virtual character	Positive emotions: ecstasy, joy, trust, acceptance, interest; negative emotions: rage, annoyance, sadness, pensiveness, apprehension — Virtual character
2018	Dimitrios Kotzias, Moshe Lichman, and Padharic Smyth	Unknown	Predicting consumption patterns with repeated and novel events	Real-word dataset, location data, social media data, music listening data — Expectation-Maximization (EM) algorithm to learn the global and individual mixing weights
2018	Chunhui Wang, Shanguang Chen, Yuqing Liu, Dongmei Wang, Shoupeng Huang and Yu Tian	Astronaut Modeling and Simulation System (AMSS)	Modeling and simulating of astronaut's performance	Human characteristic models (the changes of astronauts during spaceflight), the behavioral models (cognitive and biomechanical) and the performance evaluation models — Astronaut modeling and simulation system

Finally, several techniques and resources were used for user modeling, such as Bayesian network engine, data ontology, cognitive maps, intelligent agents, EMG sensor, website, repository and simulation systems.

5. Conclusions

Currently, an area of great interest in the design of virtual environments is the modeling of users. The basic idea of user modeling is the characterization of user behavior, that is, the personalization and adaptation of resources, processes or contents in virtual environments. In this sense, the modeling of preferences, emotions, customs, objectives, plans, characteristics of a cognitive or perceptual type, limitations, or any information that identifies the behavior of users, have been specially studied, to plan and adapt environments according to the purpose, for which they are created.

The adaptation of virtual environments according to the information of each user, allows the implementation of automatic modeling of user interactions with virtual systems. This, in turn, allows users to perform tasks in different ways: cooperative or group, individual or under a certain role.

This indicates the importance of modeling of the user lies mainly in the utility of designing environments, which can know what the user will want from the system and thus be able to orient it towards their needs and desires.

In relation to what is presented in this chapter, we believe that user modeling should be carried out as reliably as possible, that is, as real as possible in order to obtain the best results with respect to the use of the environment.

Opposite to the perceived reality, we also recognize that the user modeling is evolutionary. It is evolutionary due to the need to inquire from the user the information until reaching the final characterization within the system. The development of a user model starts with the clearer information. Subsequently the environment is evolved while incorporating new features proposed by the user or designer.

Undoubtedly, users play an important role when developing a virtual environment. This allows us to extend the normal models of design of the environments, by models that incorporate own and particular elements of each user.

In summary, within this chapter we show how user modeling, in recent years, has been used under several approaches and in different fields. Likewise, the aim of studying, analyzing and solving different problems through personalization of users in virtual environments has been very useful.

References

Aballay, L.N., Aciar, S.V., Collazos, C.A. and González, C.S. (2015). Adaptation model content based in cultural profile into learning environment.

Abel, F., Gao, Q., Houben, G.-J. and Tao, K. (2011). Analyzing User Modeling on Twitter for Personalized News Recommendations. Paper presented at the International Conference on User Modeling, Adaptation, and Personalization.

Barla, M. (2010). Towards Social-based User Modeling and Personalization. Slovak University of Technology in Bratislava, Slovak (FIIT-10890-3653).

Clemente, J., Ramírez, J. and De Antonio, A. (2014). Applying a student modeling with non-monotonic diagnosis to Intelligent Virtual Environment for training/instruction. Expert Systems with Applications, 41(2): 508–520.

Conati, C. and Maclaren, H. (2009, June). Modeling User Effect from Causes and Effects. Paper Presented at the International Conference on User Modeling, Adaptation, and Personalization.

De Rosis, F., Novielli, N., Carofiglio, V., Cavalluzzi, A. and De Carolis, B. (2006). User modeling and adaptation in health promotion dialogues with an animated character. Journal of Biomedical Informatics, 39(5): 514–531.

Emond, B., Fournier, H. and Lapointe, J.-F. (2010, April). Applying Advanced User Models and Input Technologies to Augment Military Simulation-Based Training. Paper Presented at the Proceedings of the 2010 Spring Simulation Multiconference.

Ferreira-Satler, M., Menéndez, V.H., Romero, F.P., Zapata, A. and Prieto, M.E. (2010). Ontologías borrosas para resprentar perfiles de usuario en una herramienta de gestión de objetos de aprendizah. Paper Presented at the Actas del XV Congreso español sobre Tecnologías y Lógica Fuzzy, ESTYLF.

Fischer, G. (2001). User modeling in human-computer interaction. User Modeling and User-adapted Interaction, 11: 1–2.

Frias-Martinez, E., Magoulas, G., Chen, S. and Macredie, R. (2005). Modeling human behavior in user-adaptive systems: Recent advances using soft computing techniques. Expert Systems with Applications, 29(2): 320–329.

Gascueña, J.M., Fernández-Caballero, A. and González, P. (2005). Ontologías del modelo del alumno y del modelo del dominio en sistemas de aprendizaje adaptativos y colaborativos. Paper Presented at the VI Congreso Interacción Persona Ordenador, Universidad de Granada.

Godoy, D. and Amandi, A. (2006). Modeling user interests by conceptual clustering. Information Systems, 31(4-5): 247–265.

González, G. (2014). Ontología del perfil de usuario para personalización de sistemas de u-learning universitarios. Paper Presented at the XLIII Jornadas Argentinas de Informática e Investigación Operativa (43JAIIO)-XVII, Concurso de Trabajos Estudiantiles.

Guojiang, W. and Liu, J. (2017, December). Emotional Modeling for a Vision-Based Virtual Character. Paper Presented at the Computer and Communications (ICCC), 2017 3rd IEEE International Conference on.

Hassan Montero, Y. and Martín Fernández, F.J. (2004). Propuesta de adaptación de la metología de diseño centrado en el usuario para el desarrollo de sitios web accesibles. Revista Española de Documentación Científica, 27(3).

Kaklanis, N., Moschonas, P., Moustakas, K. and Tzovaras, D. (2011, July). A Framework for Automatic Simulated Accessibility Assessment in Virtual Environments. Paper Presented at the International Conference on Digital Human Modeling.

Kobsa, A. (2001). Generic User Modeling Systems. User Modeling and User-Adapted Interaction, II, Springer, 49–63.

Kotzias, D., Lichman, M. and Smyth, P. (2018). Predicting consumption patterns with repeated and novel events. Transactions on Knowledge and Data Engineering, IEEE, 3(8): 1–14.

Lara, G. (2016). Computational Model for the Generation of Directions for Object Location in Virtual Environments: Spatial and Perceptual Aspects (Ph.D.). Universidad Politécnica de Madrid, Madrid, España.

Lasierra, N., López, M., García, J. and Alesanco, A. (2009). Desarrollo de un agente SNMP v3 para modelado de usuario en entornos LAN. Paper Presented at the VIII Jornadas de Ingeniería Telemática JITEL.

Lozano, M. and Calderón, C. (2004). Entornos virtuales 3D clásicos e inteligentes: Hacia un nuevo marco de simulación para aplicaciones gráficas 3D interactivas. Inteligencia Artificial, Revista Iberoamericana de Inteligencia Artificial, 8(23).

Moreno, L., Martinez, P. and Ruíz, B. (2008). Aplicación de técnicas de usabilidad con inclusión en la Fase de Análisis de Requisitos, (12), Departamento de Informática, Universidad Carlos III de Madrid.

Moschonas, P., Kaklanis, N. and Tzovaras, D. (2011, December). Novel Human Factors for Ergonomy Evaluation in Virtual Environments Using Virtual User Models. Paper Presented at the Proceedings of the 10th International Conference on Virtual Reality Continuum and Its Applications in Industry.

Octavia, J.R., Raymaekers, C. and Coninx, K. (2011). Adaptation in virtual environments: Conceptual framework and user models. Multimedia Tools and Applications, 54(1): 121–142.

Ortony, A. and Terence, J.T. (1990). What's basic about basic emotions? Psychological Review, 97(3): 315–331.

Peña, A.A. (2007). Un modelo del estudiante basado en mapas cognitivos. Instituto Politécnico Nacional, Ph.D. Thesis.

Quiroz, J.S., Serrano, E.F. and Cavieres, A.A. (2015). Un modelo para el diseño de entornos virtuales de aprendizaje centrados en las E-actividades Juan. Paper Presented at the XX Congreso Internacional de Informática Educativa, Santiago, Chile.

Santos Jr, E., Nguyen, H., Zhao, Q. and Wang, H. (2003). User Modelling for Intent Prediction in Information Analysis. Paper Presented at the Proceedings of the Human Factors and Ergonomics Society Annual Meeting, CA: Los Angeles.

Schoefegger, K.S., Paul and Ley, T. (2010). Towards a user model for personalized recommendations in work-integrated learning: A report on an experimental study with a collaborative tagging system. Elsevier, Procedia Computer Science, 1(2): 2829–2838.

Silva, M.D.M.S. (2016). Bases psicoeducativas en el diseño de cursos implementados en IMS-LD para el aprendizaje virtual de alumnos con diversidad funcional (Ph.D.), Universidad de Extremadura.

Skillen, K.L., Chen, L., Nugent, C.D., Donnelly, M.P., Burns, W. and Solheim, I. (2012, December). Ontological User Profile Modeling for Context-Aware Application Personalization. Paper Presented at the International Conference on Ubiquitous Computing and Ambient Intelligence.

Uribe-Rios, M.Y., Meneses-Ortegón, J.P., Jové, T. and Fabregat, R. (2016). Modelo de perfiles de adaptación en el proceso de co-creación de material para estudiantes con altas capacidades. Revista Ingeniería e Innovación Volumen, 4(1): 1–13.

Viviani, M., Bennani, N. and Egyed-Zsigmond, E. (2010, August). A Survey on User Modeling in Multi-Application Environments. Paper Presented at the Advances in Human-Oriented and Personalized Mechanisms, Technologies and Services (CENTRIC), 2010, Third International Conference.

Wang, C., Chen, S., Liu, Y., Wang, D., Huang, S. and Tian, Y. (2018, July). Modeling and Simulating Astronaut's Performance in a Three-Level Architecture. Paper Presented at the International Conference on Engineering Psychology and Cognitive Ergonomics.

Chapter **4**

Synthetic Perception and Decision-making for Autonomous Virtual Humans in Virtual Reality Applications

Hector Rafael Orozco Aguirre,[1,*] *Daniel Thalmann*[2] and *Felix Francisco Ramos Corchado*[3]

1. Introduction

Virtual humans are represented as intelligent embodied agents living in virtual environments in the form of virtual characters that look like, act like, and even interact with humans. The creation and usage of virtual humans have served to explore research issues where it is necessary to achieve cognitive systems with performance at the human level. These issues are described in (Swartout et al., 2006) and cover manifold topics in artificial intelligence. These topics span voice recognition, understanding and natural language generation, dialogue modeling, non-verbal communication, task modeling, social reasoning, and emotion and personality modeling. The animation of realistic intelligent virtual humans is a major challenge in virtual reality and artificial intelligence fields (Ali and Nasser, 2017). Current applications are now based on modeling and creating autonomous virtual humans with human-like characteristics and appearance that allow endowing them with different cognitive skills with the aim of simulating in a better way the human behavior and cognition (Newell, 1994). The virtual human technology is typically conceived as a set of practical tools to assist in a wide range of applications, such as training and entertainment. This technology is gaining interest as a methodology for studying human cognition and validates cognitive theories. By putting together multiple cognitive capabilities, such as language, gesture, emotion among others, like

[1] Autonomous University of Mexico State, Mexico.
[2] École Polytechnique Fédérale de Lausanne, Switzerland.
[3] Center for Research and Advanced Studies of the National Polytechnic Institute, Mexico.
* Corresponding author: hrorozcoa@uaemex.mx

control problems associated with navigating and interacting within a virtual environment (Gratch et al., 2013).

Cognitive architectures are a part of research in general artificial intelligence with the goal of creating programs that could reason about problems across different domains, develop insights, adapt to new situations and reflect on themselves. Similarly, the ultimate goal of research in cognitive architectures is to achieve human-level artificial intelligence. However, with no clear definition and general theory of cognition, each architecture was based on a different set of premises and assumptions, making comparison and evaluation difficult. In 2016, a broad overview of the last 40 years of research in cognitive architectures, with an emphasis on perception, attention and practical applications, was provided in Kotseruba et al. (2016). According to Lieto et al. (2018), the main role of cognitive architectures in artificial intelligence is that one of enabling the realization of artificial systems to exhibit intelligent behavior in a general setting, through a detailed analogy with constitutive and developmental functioning and mechanisms underlying human cognition. Since the last four decades, this type of application has been the focus of attention of different disciplines, such as decision-making modeling, the simulation of personality and emotion, the recognition and generation of speech, the modeling of facial expression, body animation, among others (Magnenat-Thalmann and Thalmann, 2005). These disciplines have spawned important researches that have led to significant achievements of their own on behavioral systems for autonomous virtual humans.

Swartout et al. (2006) mention that achieving human-level intelligence in cognitive systems requires a number of core capabilities, including planning, belief representation, communication ability, emotional reasoning, and most importantly, a way to integrate these capabilities. This integration has not only raised new research issues, but has also suggested some new approaches to difficult problems. However, human behavior modeling and simulation are not simple tasks because there are some questions to be answered and some problems to be solved, such as how virtual humans need to perceive their environment to make good decisions and adapt to changes into the environment (Kim et al., 2005; Bajcsy et al., 2018). These technical issues cannot be solved by just using faster computers but can be treated as a function of the new advances and results on virtual reality (Anthes et al., 2016) and artificial intelligence (Vasant and DeMarco, 2015; Müller and Bostrom, 2016).

Although there is a long journey ahead before fully replicating or mimicking the human brain functioning, recent notable advances stand out in terms of how this organ processes information from the environment, generates new knowledge and makes decisions based on the affective state induced by personality and the perceived events or stimuli from the environment. These advances in cognitive computing show the mechanisms for endowing intelligent agents with the faculties of knowing, thinking or feeling in a human-like way (Gutierrez-Garcia and López-Neri, 2015). Decoding the human brain is perhaps the most fascinating scientific challenge in the 21st century. The Human Brain Project (Amunts et al., 2016) targets the reconstruction of the brain's multi-scale organization. It uses productive loops of experiments, medical, data, data analytics, and simulation on all levels that will eventually bridge the scales. This project as an architecture is unique, utilizing cloud-based collaboration and development platforms with databases, workflow systems, petabyte storage, and supercomputers. The project is developing toward a European research infrastructure advancing brain research, medicine, and brain-inspired information technology.

A detailed overview of the study of virtual world psychology is given by (Wu and Kraemer, 2016). With the advent of *avatar*-mediated virtual worlds, psychologists now have new questions to examine and new techniques with which to study human cognition and social interactions. This overview reviews both empirical findings and theoretical perspectives and describes the dominant theoretical frameworks used to account for virtual world psychology, followed by a comprehensive discussion of the major empirical findings to date. The empirical review concentrates on cognitive and social processes that manifest in virtual settings as well as in applied areas of behavioral research.

The decision-making process was originally coined from a cognitive perspective as a psychological issue, but now this term is being recognized by more and more researchers as an important research topic in computer science for many research fields and disciplines, such as virtual reality (Jerald, 2015) and artificial intelligence (Rosenfeld and Kraus, 2018). Decision making is a cognitive process that permits selection or taking one action among several possible alternatives. This action can result in a good or bad decision. This cognitive skill must be regarded as a continuous process integrated in the interaction of the human being with its environment to react appropriately to every kind of possible situation present into it. In human behavior simulation, it is considered as one of the most interesting and important research topics. This topic promises to offer very good and useful results in several computer applications in real time, such as human behavior analysis, urban planning, natural disaster simulation, entertainment industry, among others (Rosenfeld and Kraus, 2018).

Knowledge and reasoning are two essential elements for enabling an intelligent virtual human to achieve successful behavior and take good actions in complex and dynamic environments. With this, the knowledge-based virtual humans can react and adapt to changes in the environment by updating their relevant knowledge about the environment and themselves. Table 1 shows the most important properties that virtual humans must have to be autonomous and intelligent.

The illusion of credibility is very complex modeling in intelligent virtual humans. Newell (1994) defined some of the constraints that an ideal intelligent virtual human must satisfy to better simulate human cognitive processes. These constraints are shown in Table 2. According to the above properties and constraints, a virtual human should behave in a sufficiently credible and intelligent way; that is, the human's actions must be seen as similar to the ones of a real person. However, the current computer knowledge does not yet allow the realization of a decision-making system which is able to satisfy all the above constraints. Thus, the illusion of credibility can be addressed taking into account its subjectivity and the expectations for an application which is specific.

Table 1: Properties for autonomous virtual humans.

Property	Meaning
Appearance	Need to look like a human (body, face, skin, cloth, motions)
Intelligent	Must decide by themselves in order to satisfy their goals
Social	Can interact with other virtual humans (i.e., crow simulation)
Adaptive	Able to improve their performance over time
Reactive	Should perceive and respond to changes in their environment
Proactive	Able to exhibit different behaviors
Perceptive	Perceive the environment through sensors and act upon it through effectors

Table 2: Constraints for autonomous virtual humans to simulate human cognition and behavior.

Constraint	Meaning
Flexibility and autonomy	Be able to adapt to environmental changes
Knowledge	Partial or complete information about their environment to take decisions
Cognitive skills	Plan, learn and infer new knowledge about themselves and their environment
Self-consciousness and perception	Must be able to express themselves through their actions according to each situation in their environment
Other human characteristics	Personality and emotions have to be considered to simulate in a better way human behavior and cognitive processes

Nowadays, more researchers are shifting their focus to believable agents. Belief is a more practical concept than intelligence. Based on the analysis of existing works on belief Pan and Hamilton (2018) tried to isolate the key components of belief and define a believable virtual agent as an autonomous software agent situated in a virtual environment. In other words, it is life-like in appearance and behavior, with a clearly defined personality and distinct emotional states, driven by internal goals and beliefs, consistent in behavior, capable of interacting with its environment and other participants, aware of its surroundings, and capable of changing its behavior over time. These key components are described in Table 3 with the aim of formalizing the belief and contribute to making virtual characters believable. Pan and Hamilton (2018) also attempted to develop a computational framework, implementing this belief formalism. This framework supports the implementation of believable virtual agents for virtual worlds and game engines and is labeled I2B (interactive, intelligent, and believable). They do not attempt to develop a comprehensive general-purpose belief framework, but rather present a suggestion on how the aforementioned formalism can be practically implemented.

Having artificial agents to autonomously produce human-like behavior is one of the most ambitious original goals of artificial intelligence. The design and implementation of believable artificial agents, truly indistinguishable from humans, remains an open problem. This challenge has been typically addressed from two interrelated perspectives within cognitive science, which are mentioned in Asensio et al. (2014). The first one, psychological models of human cognition, try to explain how human behavior is produced; the second one, computational models implemented in artificial agents try to replicate to some extent human-like behavior. Pan and Hamilton (2018) propose that the biggest challenge in the field would be to build a fully interactive virtual human who can pass a virtual reality Turing test, and that this could only be achieved if psychologists, virtual reality technologists, and artificial intelligence researchers work hard and together with passion and dedication. They also pointed to a big emerging question: Is really the problem of passing the virtual reality Turing test a problem that is complete with artificial intelligence? That is a problem which cannot be solved until the full intelligence of a real human is placed into a one or a set of computer systems representing a total virtual human.

Generally, human-like behavior is difficult to both define and test. In fact, the Turing test paradigm still applies to this problem because no better alternatives have been found to characterize human behavior. In the realm of computer games, this elusive characterization might, in principle, be seen easier to define (Asensio et al., 2014). From

Table 3: Key components of belief for a virtual agent.

Key Components	Meaning
Appearance	The existence of parametric *avatars* associated with the corresponding believable virtual agents. These *avatars* are defined by their visual features, e.g., height, belly size, head size, etc.
Personality	The assumption that a personality has *n*-dimensions.
Emotional state	The assumption that the emotional state has *m*-dimensions.
Alive	The agent's ability to express the illusion of life, plus verbal and non-verbal behavior.
Illusion of being alive	By uniting situation and integration into the concept of immersion in 3D virtual environments.
Consistency	To ensure the consistency of the agent behavior over the entire range of its belief features.
Change	A learning function that allows an agent to update its belief features in response to sensing a particular environment state.
Social relationship	A function, which reflects on how the current role being assumed by an agent relates to the roles of other agents.
Awareness belief	An essential component of the belief of embodied conversational behavior. Aware of where the agent is (environment awareness), who the agent is (self-awareness), and generally how the interaction is progressing (interaction awareness).
Environment awareness	The positions of objects and *avatars* in the environment, how these evolve with time and the direction vectors associated with *avatars*.
Self-awareness	Knowing own context and state within the environment (being self-aware) is essential for a virtual agent to interact believably.
Interaction awareness	The state of an agent who is 'able to perceive important structural and/or dynamic aspects of an interaction that it observes or that it is itself engaged in'.

the point of view of cognitive science, human-level intelligence and human-like behavior can be considered as produced by several interrelated psychological processes, ranging from basic activation processes like primary motivations to complex high-level cognitive processes, such as set shifting and imitation learning. The current knowledge about these processes can be used to inspire the design of artificial cognitive architectures. Asensio et al. (2014), used three different approaches to this sort of inspiration and put them to test in an adapted version of the Turing test based in a video game (Table 3).

Designing intelligent agents that interact proficiently with people necessitates the modeling of human behavior and the prediction of their decisions. Rosenfeld and Kraus (2018) stated that understanding and predicting human decision-making are the chief concerns in multiple fields, ranging from social sciences, such as psychology and economy to neurobiology and cognitive science. Unfortunately, similar to other cognitive processes, such as creativity and emotions, only a small portion of the multitude of factors which effect human decision making is currently understood by scientists. From what scientists have already established, they argue that human decision making is influenced by a large set of factors that vary across different individuals and groups of individuals, including past experiences, decision complexity, emotions, and many cognitive biases. Finally, they explore the task of automatically predicting human decision making and its use in designing intelligent human-aware automated computer systems of varying natures, ranging from purely conflicting interaction settings (e.g., security and games) to fully cooperative interaction settings (e.g., autonomous driving and personal robotic assistants).

The main research area to be presented in this chapter is the gap between psychology and computer science on decision making as the one presented by Handel (2016) as an attempt to answer the recurring question of the virtual humans: What to do next? The answers present a view on simulation of people visiting an urban event. At each moment in time, a person has to weigh the different possible actions and make consecutive decisions. For instance, a person might be hungry or thirsty and would therefore like to go somewhere to eat or to drink, or a person might need to go to the toilet and thus go searching for the restrooms. Other possible desires might be to go dancing or to take rest due to exhaustion. The main points to be discussed are how to combine and use the latest cognitive psychology achievement and computer science to simulate human behavior in autonomous virtual humans (Pan and Hamilton, 2018).

This chapter is organized as follows: Section 2 is devoted to explaining how synthetic perception can be implemented in virtual humans. The main aspects to simulate the human decision making and the most common aspects for behavioral animation are discussed in Section 3. In Section 4, a new approach to simulate human behavior is proposed. Finally, the final remarks and conclusions are discussed in Section 5.

2. Synthetic Perception for Virtual Humans

Perception is considered as one of the most important cognitive capabilities of an entity since it determines how the entity perceives its environment. The environment can be both the real and the virtual world. Perception of virtual agents is a diverse topic, as it is a vital part of the sense-think-act cycle found in both physical and virtual modern agents (Balint and Allbeck, 2013). This cognitive capacity consists of providing profitable but realistic perception processes to intelligent virtual agents with the aim of providing a solid information base to the entities' decision-making processes (Haubrich et al., 2014). In addition, a central-agent perception process should provide a common interface for developers to retrieve data from the virtual environment. The general process is evaluated by applying it to a scenario that demonstrates its benefits. The evaluation indicates that such a simulated-perception process realistically provides a powerful instrument to improve the (perceived) realism of the simulated behavior of an intelligent virtual agent.

According to Balint and Allbeck (2013), as physical humans interact within the bounds and understanding of their environment, which can change unexpectedly, virtual humans should be able to do as well. To simulate this, an agent must be able to know and understand what is going on in their environment. El Saddik et al. (2011) establish that human senses are physiological tools for perceiving environmental information through at least five senses—sight or vision, hearing or audition, smell or olfaction, touch or taction and taste or gustation. According to them, it is believed that vision and audition convey the maximum information about an environment while the other senses are more subtle.

Due to the above facts, the characteristics of the human senses have been widely investigated and simulated over the last three decades, leading to the development of reliable multimedia systems for incorporating synthetic senses into virtual characters.

Synthetic senses provide the means for virtual characters to perceive their environment through indirect means, since they usually have no restrictions when they interrogate the state of the environment in a database of virtual environment for the location and distance of the objects. Although this method of direct access is simple and fast, it has problems of scalability and realism. Because of this, recent research efforts have focused on providing agents with their own methods of perceiving the environment.

There is not believable behavior without believable perception. Many efforts have been made in the research topic of models of perception for virtual humans. Appropriated and real-time responses of virtual humans to auditory, tactile and visual events are essential to endow them with the skill of understanding their environment (Fig. 1). The perception and cognition of a virtual human is the process of collecting information about its surrounding environment, sensing and interpreting it. The perception of elements in the environment is essential for intelligent virtual humans. This skill gives them a better awareness and useful sensorial information about what is happening and changing in the environment. This information has a direct impact on their behavior. Figure 1 shows that in order to implement believable perception, the virtual humans should have a set of artificial sensors for vision, audition, and tactile sensation. These sensors filter information from the environment and provide the knowledge the virtual humans need to make decisions and interact with the environment and each other (García-Rojas et al., 2008).

Herrero and de Antonio (2004) proposed a perceptual model, which introduces more coherence between intelligent virtual-agent perception and human-being perception. This will increment the psychological coherence between the real life and the virtual environment experience. They argue that this coherence is especially important in order to simulate realistic situations as, for example, military training, where soldiers must be trained for living and surviving risky situations. Due to this coherence, a useful training would involve endowing soldier agents with a human-like perceptual model so that they

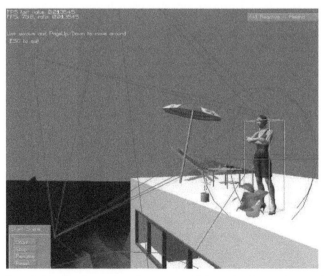

Fig. 1: Simulation of view, hearing and touch sensors with bounding boxes (*Source*: Garcia-Rojas et al., 2008).

would react to the same stimuli as a human soldier. In order to allow virtual humans to understand their environment, a generalized framework for agent sensing and perception is proposed and was implemented by Balint and Albeck (2013). This framework includes: a generalized agent-sensing system that allows simulation authors to create specific senses for a given simulation, a perception system that uses heuristics to create sense attention through a bottom-up and top-down process, and a linear combination of forms of perception that allow agents to combine perception scores across multiple senses.

By using artificial sensors, a virtual human can perceive its environment in a very similar way as a real person. However, on occasions it is not really necessary or convenient to implement artificial sensors with a high biological accuracy because it can be very time consuming and finally, inefficient. That is to say, a virtual human can perceive the environment completely at any time, but if the environment is big and complex, there could be too much information to be processed realistically in real-time. Next, the main mechanisms to implement virtual or artificial sensors are discussed.

2.1 Synthetic Vision

Human vision is considered the most important of all the senses and this has led to the sense to be mostly synthetically simulated across different domains. The aim of the different vision approaches is not necessarily to imitate the human visual system as accurately as possible; instead, it is to provide a reasonable estimate of what the visual system senses, without incurring the costs associated with simulating the highly complicated mechanisms of the human eye. According to Peters and O'Sullivan (2002), there are a number of good reasons to use an artificial vision technique. First, it can be the simplest and fastest way to extract useful information from the virtual environment; second, synthetic vision can scale better than other techniques in complex environments; third, the approach makes the virtual actors less dependent on the underlying implementation of the environment, since it is not necessary to rely directly on a database for the scenes in the virtual environment. However, research on synthetic sensors for other modalities has also been carried out.

Vision is the most important and used virtual sensor that is implemented in autonomous virtual humans, in at least to different kinds of approaches. The first one is called sensory omniscience, and the last one is known as sensory honesty. Sensory omniscience methods access directly the scene database and the virtual human becomes aware of all the objects on the scene, even those that are behind its back. This kind of vision is easier to implement and less time-consuming. But on occasions, it can produce unrealistic behaviors. On the other hand, the sensory honesty methods try to acquire the information that can actually be seen by the virtual human, as if it was a real human. These methods lead to more realistic interactions because they allow a virtual human to potentially see through difficult places, such as walls or the back of its head. These interactions are based on two main mechanisms: artificial vision and synthetic vision. There is a distinguishable difference between these mechanisms. Synthetic vision is simulated vision for digital actors while artificial vision is the process of recognizing the image of a real environment captured by a camera. Since artificial vision must obtain all of its information from the vision sensor, the task becomes more difficult, involving time-consuming tasks, such as image segmentation, recognition, and interpretation.

The concept of synthetic vision, introduced by Renault et al. (1990), corresponds to the simulated vision of a virtual human shown by a camera located on the virtual human point of view. This method combines the identification of visible objects from the scene, rendering through the virtual human point of view, with the query of the scene database to obtain the position and semantic information about these visible objects. Thus, it may be the simplest and fastest way to extract useful information from the environment, since synthetic vision could be modeled by using a Z-buffered color image representing the vision of a virtual human (Peters and O'Sullivan, 2002). In that way, each pixel of the vision input has the semantic information giving the object projected on this pixel, and numerical information to know the distance to this object. With this, for a virtual human, it is very easy to know when a particular object, like a chair, is just in front of it at one-meter distance.

Rabie (2002) proposed a paradigm for active vision research, which used photorealistic binocular retinal image streams to achieve active perception in dynamic virtual environments. Renault et al. (1990) presented a new synthetic vision mechanism based on false colors. In this technique, the objects on the scene are not rendered using their usual colors; they are rendered using a unique color for each object. The scene is rendered from the point of view of the actor and the output is stored in a 2D array containing the pixel value at that point, the distance from the eye of the actor to the pixel, and an object identifier of any object at that position. The size of the array is chosen to provide accuracy without consuming too much CPU time. A 30 × 30 matrix was used to apply this technique to avoid obstacles in a corridor.

Noser and Thalmann (1995) used a 50 × 50 matrix and improved this technique, including a visual memory model implemented with a 3D octree grid. Kuffner and Latombe (1999a) applied false color coding with a rendered image of 200 × 200 pixels. It was assumed that the environment was broken up into a collection of small to medium-sized objects, each assigned a unique ID. Large objects, such as walls or floors, were subdivided and each piece was assigned an ID. Each character maintained a kind of spatial memory, storing a list of object IDs currently visible. In Peters and O'Sullivan (2002) proposed to extend the false-colored rendering method by providing different vision modes that use different color palettes: distinct mode and grouped mode. In the distinct vision mode, each object is false-colored with a unique color; in the grouped vision mode, groups of objects share the same color. Objects can be grouped according to different criteria. Group vision mode is less precise as a group of objects can be marked as visible if at least one of its objects is visible. In this case the rendered image is of 128 × 128 matrix.

Vosinakis and Panayiotopoulos (2005) developed a different vision mechanism based on ray casting. A number of rays are cast on to the scene inside the field of view of the agent. The agent is able to know the position, size and type of objects that are intersected by the rays. Artificial vision is the process of recognizing the image of the real environment as captured by a camera (Noser and Thalmann, 1995). For example, all the information obtained by the vision sensor about the world is extracted by using 2D image analysis. This process requires time-consuming tasks, such as image segmentation, recognition, and interpretation (Peters and O'Sullivan, 2002). That is why this kind of technique is better for robotic research areas.

According to the state-of-the-art report given by Ruhland et al. (2014), eyes are central in conveying emotional information, and humans are able to interpret the intentions and feelings of other humans by observing their eyes. This fact is linked to the phrase—the face is the portrait of the mind, the eyes, its informers. This presents a huge challenge to computer graphics researchers in the generation of artificial entities that aim to replicate the movement and appearance of the human eye, which is so important in human-human interactions. This state-of-the-art report provides an overview of the efforts made in tackling this challenging task. As with many topics in computer graphics, a cross-disciplinary approach is required to fully understand the working of the eye in the transmission of information to the user. This report discusses the movement of the eyeballs, eyelids, and the head from a physiological perspective and how these movements can be modeled, rendered and animated in computer graphics applications. In Fig. 2, a virtual character is delivering 3-minute lectures to a user. Assessment included both subjective questions about rapport with the agent and objective questions about how much of the lecture was remembered.

Simsensei (Swartout, 2016) is a virtual human designed to act like an intake nurse, interviewing patients about depression. The Simsensei virtual human, Ellie (Fig. 3) uses language and non-verbal gestures such as head nods, mirroring gestures and body posture to engage and build rapport with the patient. Simsensei uses the MultiSense framework (DeVault et al., 2014), which employs machine learning to form hypotheses about the patient's condition by integrating information from multiple data streams, such as data about the patient's facial expression, body posture and activity, voice prosody, and speech content.

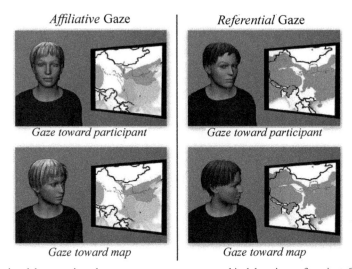

Fig. 2: A virtual human gives lectures to a user on geographical locations of ancient China in an educational scenario (*Source*: Ruhland et al., 2014).

2.2 Gaze and Attention

The modeling of interactive virtual human beings has been one of the main objectives of immersive virtual environments for training. An important characteristic of a virtual human is the ability to direct his perceptual attention to objects and locations in a virtual environment in a way that seems credible and serves a functional purpose. It is commonly seen that the amount of information in the virtual environment far exceeds the virtual human processing capabilities. In fact, only a small fraction of sensory information can be fully processed and assimilated in the cognitive model. A successful model of perceptual attention provides a path to both the elimination of

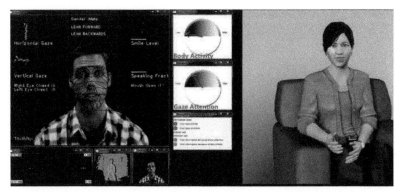

Fig. 3: Virtual interviewer Ellie (right) and MultiSense perception dashboard (left). A virtual human able to perceive and process data about a patient while helping him to feel comfortable in talking about himself (*Source*: Swartout, 2016).

Fig. 4: Mission rehearsal exercise simulation in which an injured boy is being attended to by his mother and a medico, while a sergeant is conversing with a human participant (*Source*: Kim et al., 2005).

incoming sensory information and the choice of the most relevant information to focus on during the next phase or step of a decision-making cycle. Kim et al. (2005) proposed a prototype of computational model for controlling the focus of perceptual attention for virtual humans, providing the potential to support multi-party dialogues in a virtual world. They demonstrated that virtual humans can respond dynamically to events that are not relevant to the tasks and shift their attention among objects in the environment and have given positive feedback to informal demonstrations. In order to illustrate this idea, they conducted a mission-rehearsal exercise simulation as shown in Fig. 4. An injured boy is being attended to by his mother and a medic. A sergeant is conversing with a human participant. Since the mother, the boy, and the medico are out of the visual field of view of the sergeant while the sergeant is conversing with the human, the sergeant's uncertainty levels about each of these characters will increase with time.

An important aspect which can greatly enhance the realism for virtual humans is to be aware of their environment and of other virtual entities or characters. This can partly be achieved with navigation and path planning. However, it is necessary to simulate more advanced behaviors than navigation can provide. To add attention behaviors to virtual humans (Fig. 5), two main problems are confronted—the first one is to detect the points of interest for virtual humans to look at; the second is to edit the motions for them to perform the gaze behavior. There are different models of synthetic vision. Some of them are based on memory and have been developed for the navigation of virtual humans

Fig. 5: Examples of attention behaviors in a crowd animation (*Source:* Grillon and Thalmann, 2009).

in complex virtual environments (Kuffner and Latombe, 1999b). These models were developed to simulate the human vision, but not the actual human-gaze behavior.

A model of perception was introduced by Hill (1999) and in which a character decided to attend to objects in an environment, depending on the information received from them. Khullar and Badler (2001) proposed an architecture which determined where an agent should look by selecting from top-down, bottom-up, and idling behaviors. However, their system requires that an animator insert the top-down interest points in a queue. Similarly, much work has been conducted in the simulation of visual attention and gaze in embodied conversational agents (Gillies, 2001; Peters et al., 2005; Gu and Badler, 2006). These models give very convincing results but are not applicable to crowds. Several researchers proposed perceptual systems based on saliency maps as the ones done by Itti et al. (2003) and Peters and O'Sullivan (2003). Kim et al. (2005) expanded the approach by using a benefit-and-cost function to determine when a character should look at an object. The saliency-map method gives very good results, but is prohibited for crowd animation.

Yu and Terzopoulos (2007) proposed a decision-network framework to simulate how people make decisions on what to attend to and how to react. Their system, however, is aimed at modeling and representing situations with a small group of people. Grillon and Thalmann (2009) presented a novel method to enhance crowd animation realism by adding attention behaviors to virtual humans. Here, an automatic interest-point detection algorithm is used to determine, for each character, where and when it should look. Additionally, it is presented as an extensible and flexible set of criteria to determine interest points in a scene and a method to combine them. This method also allows the fine-tuning of attention behaviors for virtual humans by introducing an attention parameter as well as the possibility to modify the relative importance of each criterion, if desired. As a second contribution, a robust and very fast dedicated gaze inverse kinematics solver to edit the character motions was proposed. After acquiring the virtual objects in sight and the audition information, the virtual humans had to limit their focus to restricted objects or entities which they were most interested in. For example, using perceptual attention, the scales can be balanced between the output of perception processing and the input of behavior decision processing. In addition, different intentions can be used to guide virtual humans to choose the objects they focus on.

2.3 Synthetic Audition

In real life, human behavior is very often influenced by sounds. Audition is a temporal sense and humans are very sensitive to changes in acoustic signals. Humans can locate objects in space, especially when they are moving. Acoustic signals transport a lot of semantic and emotional information and tell them a lot about the position of a sound source relative to them and the propagation paths in the environment. At the moment, research on perception models for intelligent virtual agents is focused more on visual perception than on auditory or hearing perception. Most of the studies carried on perception in virtual humans focus on visual perception. Although vision is the most analyzed sense, a virtual human not only can perceive information from the environment through visual sensors, but also by auditory sensors (Noser and Thalmann, 1995).

Important contributions in the field of auditory virtual environments have been made by Blauert (1974) in describing the psychophysics of human-sound localization. Other key features are the introduction of physics and perception of sound and digital signal processing related to spatial hearing and the outline of the components of spatial auditory displays. Noser and Thalmann (1995) presented a simple real-time structured sound renderer. This sound renderer is used as an audition channel for synthetic and real actors and synchronized sound track generator for video film productions. The virtual audition information can be sensed by virtual humans through interrogation of system messages. To simulate the transmission feature of vocal information, after checking the system message, each virtual human should check the distance scalar between the position where the message is given and its own position in order to judge if the vocal information can be heard by it.

There are many factors that contribute to the human capacity to perceive sounds coming from the environment and which cannot be easily modeled or reproduced in a virtual environment. An analysis of some key concepts of human auditory perception to be introduced in a perceptual model of auditory agents was carried out by Herrero and de Antonio (2003) with the aim of making a more human-like perceptual model. A case study to test this perceptual model can be appreciated in Fig. 6. The concepts, selected for being the most representative of human auditory perception, were auditory acuity, directivity of sound, inter-aural differences, auditory filter and, finally, cone of confusion.

A perception system that closely models human sensory systems is critical for simulating human-like agents evolving in a dynamic, non-deterministic environment. Steel et al. (2010) proposed a modular multi-sense perception system for virtual agents. This system is extensible and modifiable by providing a plug-in interface for perception algorithms, allowing users to dynamically modify how an agent perceives its environment. The perception module disseminates information to its various sensors,

Fig. 6: The war scenario to endow soldier agents with a human-like perception model (*Source*: Horrero and de Antonio, 2003).

providing details for only visual and auditory sensors. Although some works have been published on auditory perception, olfactory perception seems to have been left aside by most of the researchers working on synthetic perception. But Kuiper and Wenkstern (2013) discuss a combination of perception algorithms for virtual agents. In this way, agents are not provided with global knowledge, but acquire environmental knowledge through their perception module that integrates several sensors. Each sensor can be configured individually for each agent and can be modified at runtime. The combination of perception algorithms mixes vision, auditory and olfactory sensory data to produce knowledge.

2.4 Synthetic Tactile

One important characteristic to be added to virtual humans is to endow them with a tactile sensory skill to mainly interact with objects in their environment. This sensory ability can be used in tasks, as touching or moving objects. The use of tactile sensors provides the virtual humans physical details about the objects in their environment. In real life, humans can sense physical objects if any part of the body touches them and gather sensory information. This sensory information is made use of in such tasks as reaching out for an object, navigation, etc. For example, if a person is standing, the feet are in constant contact with the supporting floor. But during walking motion, each foot alternately experiences the loss of this contact. Traditionally these motions are simulated, using dynamic and kinematic constraints on joints. Nevertheless, there are some situations where information from external environment is needed. One of them is when a person descends on a staircase. In this case, the motion should change from a walk to descent, based on achieving contact with the steps of the stairway. Thus the environment imposes constraints on human locomotion and this last one must be according to the objects and terrain that are being touched.

The necessity to model interactions between objects and agents like virtual humans appears in most applications of computer animation and simulation. An example of an application using agent-object interactions is presented by Johnson and Rickel (1997) and whose purpose is to train equipment usage in a populated virtual environment. Another interesting way is to model general agent-object interactions based on objects containing interaction information of various kinds: intrinsic object properties, information on how to interact with it, object behaviors, and also expected agent behaviors.

As already mentioned, simulating tactile sensors corresponds roughly to a collision detection process. In order to simulate sensorial tactile events in Thalmann et al. (1997), a module was designed to define a set of solid objects and a set of sensor points attached to a virtual human or actor. The sensor points can move around in space and collide with the above-mentioned solid objects. Collisions with other objects out of this set are not detected. The only objective of collision detection is to inform the actor that there is a contact detected with an object and which object it is. Standard collision detection tests rely on bounding boxes or bounding spheres for efficient simulation of object interactions. But when highly irregular objects are present, such tests are bound to be ineffective. Feng et al. (2012) presented a highly skilled interactive virtual character by combining locomotion, path finding, object manipulation, and gaze. They synthesize natural-looking locomotion, reaching and grasping for a virtual character in order to accomplish a wide range of movement and manipulation tasks in real time. Virtual characters can move while avoiding obstacles, as well as manipulate arbitrarily-shaped objects, regardless of

Fig. 7: A virtual human grasping objects with various shapes. The virtual human is able to determine hand orientations and finger poses based on object shapes and collision detections (*Source*: Feng et al., 2012).

height, location or placement in a virtual environment. Characters can touch, reach, and grasp objects while maintaining a high-quality appearance. They demonstrate a system that combines these skills in an interactive setting suitable for interactive games and simulations.

A virtual human must be able to reason out what it perceives from its environment to make the best decisions in order to satisfy its goals and survive in the changes in the virtual world. This means that the virtual human should react to every situation that takes place in its environment and decide its actions by reasoning out what it knows to be true at a specific time. With this, its knowledge is decomposed into beliefs, desires, and intentions to formulate new plans or sequences of actions required to achieve its goals. The combination of perceptions of virtual agents is still a new and challenging research area, with great potential for new ideas. Figure 7 shows a character animation system that integrates different controllers to achieve a reaching task. By utilizing the path planner, motion blending, and inverse kinematics, the system enables a virtual character with skills to maneuver obstacles and precisely grasp a target object in space. The system can respond to dynamic environments to generate a new reaching motion in real-time and is therefore suitable for interactive applications, like video games. But collision avoidance to handle obstacles during arm movements is missing.

2.5 *Multisensory Perception and Multi-party Interaction*

Current trends in recent research on virtual humans focus on multisensory perception. A good example for a virtual human with this kind of perception is shown in Fig. 8. This virtual human, called Chloe, is able to replace a person absent from a meeting, or could play a dedicated role, such as meeting a leader or a virtual secretary. In order to achieve a realistic behavior, the virtual human is able to detect visual and audio cues, managing emotions, and to recognize speech from users. This virtual human was created in 2012 in the Being There Centre of the Institute for Media Innovation (today named as Being Together Centre) at the Nanyang Technological University in Singapore, which is a result of the integration of research conducted by different researchers. In order to have as close as possible a real-life human-to-human interaction, some key features and capabilities for Chloe are the following ones:

- Real-time face recognition system, using webcam, enables Chloe to identify people to whom it is talking

Fig. 8: Chloe, a virtual human, interacting with a user in a human-to-human-like interaction.

- Head and face tracking gives Chloe the ability to look at people and follow them with its gaze
- Chloe can recognize the gestures of the person talking to it, with the aid of the Kinect sensor that provides it a sense of depth and motion
- Memory management provides Chloe with basic knowledge and helps to store information about past encounters
- Emotion and personality models give Chloe a more natural behavior that will adjust along with the conversation flow
- An animation system controls Chloe's movement and when synthesized with the speech and animation of its lips, gives it the ability to speak
- Speech recognition enables Chloe to have a more natural human-to-human interaction

A gaze behavior model for an interactive virtual character, called Sara, situated in the real world was recently proposed by Yumak et al. (2017). This virtual human can interact with two users at the same time as shown in Fig. 9. In this model, they were interested in estimating which user has the intention to interact; in other words, which user is engaged with the virtual human. The model takes into account behavioral cues, such as proximity, velocity, posture, and sound, estimates the engagement score, and drives the gaze behavior of the virtual character. However, sound, vertical head orientation, and field of view parameters didn't behave as was initially expected.

The other trend is related to multi-party interaction between human beings, robots and virtual humans. Yumak et al. (2014) are in particular focusing on multi-modal and multi-party interactions with a virtual character, a human-like social robot and two human users and provide a model of multiparty interaction to explore the dynamics of multi-party interactions in natural settings (Fig. 10). In this case, the authors said that both the virtual human and the robot can interact with the users and engage in a multi-party conversation, which brings numerous challenges as more than one person can engage with the ongoing conversation, participate, and leave any time. This requires that each agent will keep track of the actions of other participants. In order to interpret the users' actions, they need to understand the scene as a whole by fusing information from different channels of information. This research aims to be part of a bigger vision where

Fig. 9: Two users engaged with a virtual human are able to drive its gaze behavior (*Source*: Yumak et al., 2017).

Fig. 10: Multi-party interaction among humans, a virtual human and a human-like social robot (*Source*: Yumak et al., 2014).

the virtual characters, social robots, and human beings live and interact together in the future, leading to cyber-societies.

The next section gives a review of the main aspects used to simulate human behavior.

3. Behavioral Animation

When there is concern in animating virtual humans, it can be seen that there is a huge difference between guided behavior and autonomy. When a virtual human is guided, it will do what the animator specifies. To do that, the animator must predefine its behavior, which results in appropriate behavior and thereby cannot be used in applications where a dynamic interaction between virtual humans is required. On the contrary, if a virtual human is autonomous, the task for the animator is minimal because it is only necessary to provide a 3D model. But for the programmer, it could be time consuming to develop good algorithms. These algorithms must provide an autonomous behavior and a dynamic animation for the virtual human and the final results are not always the best ones.

Behavioral animation of virtual humans consists of modeling their behavior, rather than being focused only on their physical appearance or features. In this kind of animation, the virtual humans can be autonomous and decide their actions. This fact gives them the ability to improvise, and frees the animator from the need to specify each detail of every motion and behavior. This section focuses on the action-selection aspect of behavioral modeling; that is, it is concerned with how virtual humans decide what to do at any given time. Behavioral architectures offer the best characteristics to develop autonomous virtual humans. In these architectures, it is possible to incorporate a natural mechanism of the decision-making process to solve two main problems: the action selection and learning from experience; that is to say, this mechanism must allow reinforcing of the effectiveness of existing behaviors and training to acquire new behaviors.

3.1 Learning Mechanisms

In behavioral systems, learning is generally considered from two perspectives: reinforcement of the effectiveness of an existing behavior, and training for acquiring new ones. The majority of behavioral architectures exhibit the use of adaptive agents that are able to handle unexpected situations due to their reactive abilities, but they do not learn from the feedback provided by the environment. Through the use of realistic virtual humans, many systems allow the simulation of complex behaviors, but all the functionality of agents is usually implemented manually. Introducing a learning mechanism inside a behavioral architecture has several advantages. In effect, this mechanism allows the following:

- Obtaining and maintaining robust and autonomous agents for long periods of time simulation
- Reliable mechanisms for developing autonomous agents and their behavior
- An efficient management of the functionality of the agent for avoiding a collapse and reprogramming their behavior in a changing environment

A great deal of research has been performed on the control of animated autonomous characters. The various techniques have produced impressive results, but they are limited. They have no learning ability and are therefore limited to explicit pre-specified behaviors. Online behavioral learning has begun to be explored in computer graphics. Reinforcement learning is shown to work well in Conde et al. (2003). Also, since all these learning techniques are designed to be used online, they have, for the sake of interactive speed, a restricted learning capacity. Reinforcement learning is learning through trial and error via the reception of a numerical reward. The agent attempts to map the state and action combinations to their utility, with the aim of maximizing the future reward. Reward is usually received after a number of actions have been taken by the agent in its environment. Reinforcement learning has been applied to a wide variety of domains, such as game playing, control, scheduling, and many others. Some of the earliest research in computer graphics involving reinforcement learning sought to make virtual characters automatically learn how to walk, swim, or jump. However, while interesting and useful, this type of learning does not provide them with decision-making abilities. The goal of this kind of learning is to automatically learn an optimal policy using, as information available, a fitness function. Optimal implies that the policy always maps the current state to the best possible action according to the fitness function.

Since it is impractical to store a large Q-factor table explicitly, it must be approximated. Previously, Conde et al. (2003) demonstrated the difficulty of this approach. Firstly, to obtain stable results, the Q-factor table must be approximated to a high degree of accuracy; secondly, Q-learning is harder to perform if there are no terminal states, and can be challenging to successfully use, especially since it requires the animator to visit every state-action many times. Conde and Thalmann (2006) presented a new integration

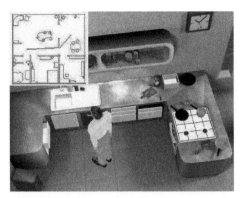

Fig. 11: Top view of a virtual apartment where a virtual human learns how to navigate into it, using path planning based on a low-level learning process (*Source*: Conde and Thalmann, 2006).

approach for simulation and behavior in the learning context that is able to coherently manage the shared virtual environment for the simulation of autonomous virtual humans (Fig. 11). This approach is a low-level learning technique that is fast, simple, and robust, and is more useful to the computer-graphics community than a technique based on the classical Q-learning approach. It is also able to automatically learn behavioral models for difficult tasks. The objective is to allow an autonomous virtual human to explore its unknown environment to build structures in the form of cognitive models or maps, based on this exploration (Fig. 11). Once its representation has been constructed, the virtual human can then easily communicate its knowledge to, for example, other virtual humans.

Simulating believable virtual crowds has been an important research topic in many research fields, such as industry films, computer games, urban engineering, and behavioral science. One of the key capabilities that agents should have is navigation, which is reaching goals without colliding with other agents or obstacles. Lee et al. (2018) presented an agent-based deep-reinforcement learning approach for navigation, where only a simple reward function enables agents to navigate in various complex scenarios. This method is also able to do that with a single unified policy for every scenario, where the scenario-specific parameter tuning is unnecessary. This agent-based approach is used for simulating virtual agents in crowded environments. Figure 12 shows two groups of agents representing pedestrian virtual humans passing and walking without colliding with each other.

3.2 Belief-Desire-Intention Models

In a virtual environment, virtual humans need to reason to take autonomous decisions. To do this, they must be endowed with a richer model of reasoning to fully understand their environment and other entities. Intelligent characters, like autonomous virtual humans, can abstractly reason using concepts, such as beliefs, goals, plans and events (Bogdanovych and Trescak, 2016). These types of characters are often referred to as Belief, Desire and Intention (BDI) agents (Meneguzzi and De Silva, 2015). Ideally, agents are able to display reactivity, proactivity, and social abilities; that is, they should be able to perceive and respond to the environment, take initiatives in order to satisfy their goals, and be capable to interact with other (possibly human) actors. For an agent to have an effective balance between proactive and reactive behavior, it should be able to reason

Fig. 12: Two groups of agents colored in white and dark gray are simulated by a policy learned by a deep-reinforcement learning approach (*Source*: Lee et al., 2018).

about a changing environment and dynamically update its goals. Having social abilities requires it to respond to other agents, for example, by cooperating, negotiating, or sharing information. Working in a team, for instance, requires agents to plan, communicate, and coordinate with each other. BDI architectures lend themselves well to implementing these requirements in an intuitive, yet formal way. These architectures have been used to design intelligent virtual humans in dynamic environments. These models offer a good methodology to create autonomous virtual humans with a complex behavior, where beliefs influence their actions. This means the virtual humans can respond flexibly to changing circumstances, despite incomplete information about the state of the environment and other virtual entities in it (Caillou et al., 2015).

Since BDI uses mental attitudes, such as beliefs and intentions, it resembles the kind of reasoning that we appear to use in our everyday lives (Harland et al., 2017). To interact successfully, agents can benefit from modeling the mental state of their environment. Additionally, BDI models provide a clear functional decomposition with clear and retractable reasoning patterns, which can be extended by including emotions to have EBDI models as shown in Fig. 13. When a new event is perceived and based on its personality and current affective state, a virtual human updates its internal beliefs (about itself) and external beliefs (about the environment) to activate new desires and generate new intentions.

The EDBI approach can provide more helpful feedback and more explanatory power in emotion-simulation scenarios for virtual humans. Emotions reflect the current affective state of virtual humans. These are updated, taking into account their personality profile. Beliefs are the knowledge available to virtual humans about themselves and their environment. Desires are the goals that the virtual humans try to achieve, while intentions are the actual actions selected by the virtual humans to be executed.

3.3 Classifier Systems

Classifier systems are the best behavioral architectures because the action-selection process is done by using the cycle of perception-decision-action. Therefore, this loop is an intuitive representation that simulates the human behavior in a more natural way; that is to say, how human beings decide what to do at any given time based on their

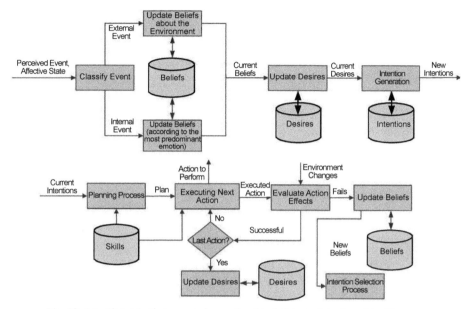

Fig. 13: Intention-selection process in emotional-belief-desire-intention virtual humans.

perception from their environment. In addition, these systems are easy to modify and extend by adding more human characteristics, such as personality (characterization of the behavioral patterns of an individual that influence its behavior), emotions, because these influence goal construction in problem solving and decision making, and also beliefs, desires, and intentions to develop a richer model of reasoning with a higher level of abstraction that permits to model human mental states (Orozco et al., 2010).

Decision making based on learning classifier systems allows creating a system which is able to manage complex situations and reactions in both static and dynamic environments. Classifier systems represent a compact and useful behavioral mechanism based on an intuitive and natural description of human behavior by using compact and expressive rules. These systems operate on the principle of loop perception-decision-action based on the mechanisms of natural selection and genetics, and include learning from the past. This permits intelligent agents, such as virtual humans, to extend their knowledge and improve their performance over time. A classifier system is a good tool that offers a suitable solution to simulate complex decision-making processes, and also provides a high flexibility in cases where a quick answer is more important to provide the best possible alternative to the current situation. However, in some cases, supervision is still required to help the system to learn about the best solution according to the situation and events that are happening in the environment. This supervision can be automatic if there are simple rules to simulate the behavior, or using human supervision for complex cases. In addition, contrary to rule-based systems that cannot provide a response to an unknown situation-event pair, a learning classifier system is able to provide a fast solution and apply a reward policy as a feedback mechanism to adapt this response for future use. Finally, basic knowledge can be introduced into the system to provide a predefined behavior in cases where the results are necessary.

A classifier system represents a type of rule-based behavioral architecture that uses an evolutionary genetic algorithm on the basis of production rules (classifiers) to identify and reinforce a subset of rules that are able to cooperate to achieve a specific task. Classifier systems are an attractive alternative for human behavior simulation in intelligent agents as virtual humans because they respect most of the constraints expressed in Table 2. These systems have the following characteristics:

- A rule base is used to represent the knowledge of the virtual human. These rules have the form *condition::action*. The representation of these rules permits to symbolically express the different environmental situations and associate each of these situations to a pertinent action.
- The system operates on the principle of the loop perception-decision-action. Thus, an action is selected, based on the rules whose condition part matches the situation perceived by the sensors and is then transmitted to the environment through the effectors.
- Generally these systems are very reactive; that is, these produce actions taking into account the perceived stimuli from the environment.
- The combination of a selection mechanism based on the relevance of the rules and a reinforcement algorithm allows the persistence and generation of behavior in a more efficient manner.
- The adaptability of the system is ensured by the application of a *covering operator* for the dynamic generation of rules, depending on the situation and a *genetic algorithm* to generate new classifiers which are potentially more efficient.
- These systems exhibit a high propensity for the generation of general rules. This allows using the system in different environments and reducing the necessary time for selecting an action based on the current situation.
- The ability of the classifier systems to work in complex environments in an efficient way, represent a compact behavioral structure based on an intuitive and natural description of the human behavior by using compact and expressive rules. However, this depends on the classifier in use.
- Finally, a classifier system is able to simulate reactive behaviors and plan long sequences of actions.

Evolutionary computing techniques are search algorithms based on the mechanisms of natural selection and genetics; that is, they apply the *principle of the survival of the fittest* among computational structures with the stochastic processes of gene mutation, recombination, etc. This principle was introduced by Darwin in his book entitled *On the Origin of Species*. Central to all evolutionary computing techniques is the idea of searching a problem space by evolving an initially random population of solutions so that better solutions are generated over time. Thus, the population of candidate solutions is seen to adapt to the problem. These techniques have been applied to a wide variety of domains, such as optimization, design, classification, control, and many others.

3.4 *The Role of Personality and Emotions*

Several considerations must be taken into account when human characteristics, such as personality and emotions, are added into virtual humans with the goal of making them more conceivable and believable. Hence, good models of behavior based on personality and emotions can help us to design and build better software agents, which approach the behavior of human beings in order to adapt and survive with more success in hostile and unpredictable environments that can change dynamically over time. Thus, it is possible to find several publications entirely dedicated to proposing different models of behavior of human beings. These models generally are used to explain human emotions. However, at this moment, it is not possible to find a universal model that describes how emotions and moods affect our behavior based on our personality and the events and stimuli we perceive from our environment. According to Saberi et al. (2014), the term 'personality' refers to consistent patterns of emotions, thinking and behavior, which make humans unique and distinguishable. During daily human-to-human interactions, people evaluate the personality of others, e.g., to predict their behavior, to understand them, to help or to motivate them. One of the important sources of information people rely on when attributing personality to others is non-verbal behavior, such as gestures, body stance, facial expressions, and gaze behavior. They develop a biologically and psychologically grounded computational architecture for generating non-verbal behavior to express personality for virtual humanoid characters. The architecture was designed in a way which can generate plausible dynamic behavior for the virtual humanoid character in response to the human user's inputs in real time.

Human modeling is an interdisciplinary field research on intelligent virtual humans as a new subject for artificial intelligence. A believable virtual human can make human-like behavior; expressing autonomous emotion in a virtual environment. Recent researches on psychology and cognitive science show that emotion is very important in human decisions. Modeling emotion is an interesting subject in artificial intelligence. The generation of autonomous behavior in virtual humans is associated with their perception and adaptation to their environment to take correct decisions. The automatic generation of behaviors in complex and dynamic virtual environments allows the virtual humans to interact more naturally with such systems, and thereby, improve their performance and efficiency to accomplish their goals. Thus, this type of behavior enables virtual humans to improve their actions, taking into account their beliefs, desires, and intentions. Research on autonomous virtual humans is a new subject for artificial intelligence. A believable virtual human can make human-like behavior, expressing autonomous emotions in a virtual environment. Recent researches on psychology and cognitive science show that emotion is very important in human decisions. Modeling emotion is an interesting subject in artificial intelligence.

A perfect emotion model should consider the human cognitive mechanism; it should model stimuli and the human mental process. How to calculate emotion intensity under a stimulus is still a question. Minski first pointed out the importance of emotions in artificial intelligence in one of the best books to be published on emotion machine (Minsky, 2006). As is cited by Steunebrink et al. in Ortony et al. (1998), the famous OCC theory of emotions providing a clear and convincing structure of the eliciting conditions of emotions and the variables that affect their intensities is presented. But this model has a number of ambiguities in the logical structure, and if these ambiguities are not resolved, computer scientists wishing to formalize or implement emotions may come up with conflicting interpretations of the psychological model. Kshirsagar and Magnenat-Thalmann (2002) give a simplified definition to say mood is a transition from a bad affective state to a good affective state. Liu and Lu (2008) presented a computational model of motivation, integrating personality, motivation, emotions, behavior and stimuli together. In spite of the fact that this model shows how motivation and personality drive the emotions of a virtual character, it only gives a primary outline of a motivation model and is restricted to be tested by a 3D-facial animation system.

A new framework based on artificial intelligence for decision making is introduced by Iglesias and Luengo (2007). This framework is used to produce animations of virtual *avatars* evolving autonomously within a 3D environment. The exposed animations in this framework are not very realistic because the *avatars* follow a behavior pattern from the point of view of a human observer. A model of individual spontaneous reactions for virtual humans is proposed by Garcia-Rojas et al. (2008). This model was defined by analyzing real people reacting to unexpected events, presenting a semantic-based methodology to compose reactive animation sequences using inverse kinematics and key frame interpolation animation techniques. Nevertheless, this model was created in a subjective way, according to the personal judgment of authors. In addition, the reaction types and animation sequences that the virtual humans perform are not validated from the point of view of psychology, but the obtained results are very satisfactory.

Another interesting approach proposing a model for conversational virtual humans that describes emotions, moods, personality and their interdependencies using vector algebra is proposed by Egges et al. (2004). In order to validate this model a 3D talking head, called Julie, which is capable of rendering speech and facial expressions in synchrony with synthetic speech, was implemented in combination with a dialogue system. They have developed a small interaction system that simulates Julie's behavior (Figs. 14 and 15). They performed these simulations with different personalities, obtaining different results in the interaction and the expressions that the face shows. The interaction that takes place for an extravert personality (90 per cent extravert) is shown in Fig. 14. Figure 15 shows how the personality and emotion-based response is generated in this case. However, when we change the personality so that it becomes more neurotic (90 per cent neurotic), the interaction also changes significantly.

Many of the revised works, which address the use of personality in the behavior of virtual creatures, make mistakes when assigning random values to the different basic personality traits. The allocation of these values cannot be supported because the used theoretical framework does not make sense from a psychological point of view, indicating that so far none of the existing works provides an accurate and reliable mechanism for modeling human behavior. Orozco et al. (2011) proposed a fuzzy mechanism to update in a more natural way the emotional and mood states of virtual humans with a real personality

\<Man\>   Can I help you carry this box upstairs?
\<Julie\> |GR30| Oh thank you! That is so nice of
you to ask!

\<Man\>   On what floor are you living?
\<Julie\> |GR99| The third floor. It's quite high!
I am so glad you are helping me out!

Fig. 14: Julie's behavior as an extravert personality (the extraversion parameter is set to 90 per cent) (*Source*: Egges et al., 2004).

<Man> Can I help you carry this box upstairs?
<Julie> |FE34| No I don't need your help

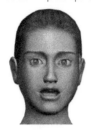

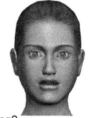

<Man> On what floor are you living?
<Julie> |FE99| That is none of your business.
Go away!

Fig. 15: Julie's behavior as a neurotic personality (the neuroticism parameter is set at 90 per cent (*Source*: Egges et al., 2004).

and emotional intelligence. The construction of a hierarchical fuzzy rule-based system to model personality and emotions for a story character is proposed by Su et al. (2007). Magnenat-Thalmann and Thalmann (2005); Burden and Savin Baden (2019) discussed several methods to model the autonomous behaviors of virtual humans. Due to the direct correspondence between emotions and facial expressions (Jiang, 2008; Ivanović et al., 2015), many researchers prefer to employ the six basic emotions of Ekman and Davidson (1994) for facial-expression classification and the OCEAN (openness, conscientiousness, extroversion, agreeableness, and neuroticism) model (Costa and McCrae, 1992), or else the OCC model (Steunebrink et al., 2009) in combination with the OCEAN model (Costa and McCrae, 1992). The mutual dependence between emotions and personality is often represented by Bayesian belief networks (Ball and Breese, 2000).

Affective systems vary, depending on their purpose and complexity. The simplest systems use emotions only as input or output, i.e., just to express emotions. However, to simulate human behavior, it is also crucial for an artificial agent to actually 'feel' the emotion meaning when performing rational processes. Pudane et al. (2017) presented a review and comparison about existing affective agent architectures, using roles of emotions as a criterion as well as proposes in architecture to create truly affective agents, i.e., agents that have full-fledged emotions. They concluded that although there are some architectures that are well developed, the field of affective computing still lacks an architecture that would fully implement human emotional capabilities. This ongoing research aims at developing intelligent agents that are able to express and incorporate affects into rational processes. Nunnari and Heloir (2017) proposed a method to generate a virtual character, whose physical attributes reflected public opinion of a given personality profile. An initial reverse correlation experiment trained a model which explained the perception of personality traits from physical attributes. The reverse model, solved by using linear programming, allows for real-time generation of virtual characters from an input personality. This method has been applied on three personality traits (dominance, trustworthiness, and agreeableness) and 14 physical attributes, verified through both an analytic test and a subjective study (Fig. 16).

4. Human Behavior Simulation

In this section, a new model is proposed to simulate more believable and realistic behaviors in virtual humans with a real personality (Sellbom, 2019) and emotional intelligence

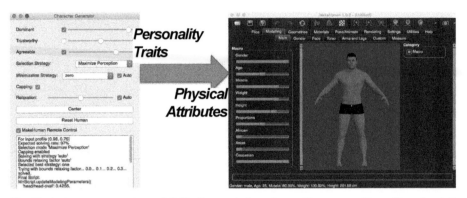

Fig. 16: A character generation tool (left) allows a designer to provide a personality profile as input. It automatically modulates physical attributes of a character editor (MakeHuman, right) so that the resulting character's appearance matches the personality profile given as input (*Source*: Nunnari and Heloir, 2017).

(Mayer et al., 2016). This model is based on the Triune Brain Model (Armstrong and Falk, 2012). A brief view of the operating cycle of the model (Fig. 17) is as follows:

- Virtual humans perceive events from their environment through sensors. These sensors transmit values (positives or negatives) of all captured events to them. Thus, once an event has been perceived, this is unconsciously processed by the reptilian brain of the virtual humans, generating an instinctive reaction that may be of two types: a *reflex reaction* or an *instinctive reaction* of protection. Once the event has been received, this unconsciously generates an affective-state (emotional and mood states) update that generates an emotional response. This update is based on the personality of the virtual humans and the level of intensity (positive or negative) of the perceived events. This process occurs in the mammalian brain of the virtual humans. Orozco et al. (2011) explain this process. Immediately, in parallel to the generated affective-state update process, the emotional response and the instinctive reaction, the virtual humans become aware of the perceived events and interpret their affective state to decide whether they have received events that catch their attention.

- Thus, the virtual humans become aware of the perceived events and search for an explanation by means of looking for information from their beliefs and long- and short-term memories. Once the virtual humans collect necessary information, they can regulate their affective state by applying their emotional intelligence to control their affective state and then to show a better behavior that is consistent with their personality. Next, the virtual humans update their desires and intentions. Based on their new desires and intentions, the virtual humans choose to execute a set of actions that are considered most appropriate to the situation. These actions are generated through dynamic planning. Finally, the virtual humans are capable of evaluating obtained results from their exhibited behavior and learn about them. For example, if certain beliefs result to be false in an experienced situation, then the virtual humans update them.

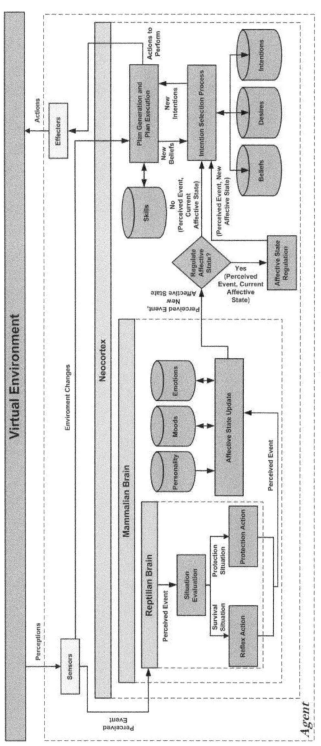

Fig. 17: Operating cycle of the proposed behavior model.

4.1 *Event Perception and Situation Evaluation Process*

Virtual humans perceive through sensors and the information provided by them is interpreted as the occurrence of a set of events. In addition, these events contain domain knowledge that provide the level of intensity (positive or negative) as well as environmental information. Then, in this way, perception represents the beginning of the primary cause of action and behavior in a virtual human. Thus, the way a virtual human perceives its environment affects its affective state, decisions, and actions to be executed in the environment according to its beliefs, desires, and intentions. When a virtual human perceives an event from its environment, this event is firstly processed by the reptilian brain in a quick and unconscious way, to generate an automatic reaction that could be either a reflex action or a protection action. Both reactions are predefined and their sensorial intensity is perceived from the captured event. The sensorial information associated with an event e, has two parameters: the level of intensity (int) and environmental information (inf). Then a definition of a captured event is denoted as follows: $e(int, inf)$. The perception of the virtual humans is represented by generation of events, whose formalization is given in event calculus (Patkos et al., 2016) as follows:

$$PERCEPTION = \sum Happens(e(int, inf), ti)$$

That is to say, perception is the sum of all events occurring in the environment.

4.2 *Intention Selection Process*

In order to represent this, a new EBDI (Emotion-Belief-Desire-Intention) architecture similar to the ones presented by Jiang et al. (2007); Bourgais et al. (2016) is proposed, considering time constraints and internal and external events that affect the affective state of virtual humans. Internal or motivational events are performed by the free will of virtual humans in pursuance of their goals. These events affect their desires and intentions and can update their beliefs. External or informational events represent sensorial information perceived from the virtual environment. These events only influence the beliefs of the virtual humans. With the aim of formalizing the EBDI-based intention selection process, the event calculus (Patkos et al., 2016) was selected by its intuitive logical definition of internal and external events and their effects over time. The formalization of events endows the virtual humans with cognitive skills to assess events and properly react to them. Next, the list of the event calculus is utilized to define how EBDI virtual humans perform actions in their environment:

- *Initiates (e(), f, t)*: A fluent f holds, after an event $e()$ is perceived at time t.
- *Terminates (e(), f, t)*: A fluent f does not hold, after an event $e()$ is perceived at time t.
- *HoldsAt (f, t)*: A fluent f holds at time t.
- *Happens (e(), t)*: An event $e()$ is perceived at time t.
- *Initially P(f)*: A fluent f holds from $t = 0$.

Fluent are variables that change over time. Herein, only Boolean fluents are considered. Different sets of fluents are defined. These are the following:

- A set of emotions $E = \{anger, disgust, fear, happiness, sadness, surprise\}$ that correspond to the six basic emotions defined by Ekman.

- A set of beliefs $B = \{b_1, b_2, \ldots, b_i\}$ representing assumptions for the virtual humans about the state of the environment and their possible skills.
- A set of desires $D = \{d_1, d_2, \ldots, d_j\}$ mapped to the goals of the virtual humans.
- A set of intentions $I = \{i_1, i_2, \ldots, i_k\}$ containing a set of committed plans formed by different actions.

Additionally, the predicate *Predominant* (E) is defined. It receives as the parameter the set of emotions E and returns the most predominant emotion $emotion_k \in E$. The result of this predicate is computed from the fuzzy evaluation of the events presented above. Assumptions about the initial state of the virtual humans are made with respect to their initial beliefs and initial desires. Intentions of the virtual humans establish the desires they want to achieve in a given moment and represent all the possible actions that a virtual human selects in order to follow a given course of action to achieve a goal, according to its affective state and beliefs. Events that happen in the virtual environment can be either external or internal. External events (*ee*) are when the virtual humans perceive changes in their environment, and internal events (*ie*) are when these refer to actions performed by the virtual humans in the achievement of their goals. External events update the beliefs of the virtual humans, without considering any emotional state. These just provide new beliefs about the virtual environment. An external event initiates a set of beliefs and terminates another set of beliefs about the environment. This is formalized as follows:

$$Initiates\ (ee(),b_1,\ t)\ ^\wedge\ Initiates\ (ee(),b_2,\ t)\ ^\wedge\ \ldots\ ^\wedge\ Initiates\ (ee(),b_i,\ t)\ ^\wedge$$
$$Terminates\ (ee(),b_{i+1},\ t)\ ^\wedge\ Terminates\ (ee(),b_{i+2},\ t)\ ^\wedge\ \ldots\ ^\wedge\ Terminates\ (ee(),b_j,\ t)$$
$$\leftarrow Happens\ (ee(),\ t)$$

Internal events are attached to the emotional state of virtual humans; that is to say, according to the most predominant emotion, virtual humans update their beliefs. For example, supposing the most predominant emotion of a virtual human is anger, this entity could believe that others are against it. This update process is represented as follows:

$$Initiates\ (ie(),b_1,\ t)\ ^\wedge\ Initiates\ (ie(),b_2,\ t)\ ^\wedge\ \ldots\ ^\wedge\ Initiates\ (ie(),b_i,\ t)\ ^\wedge$$
$$Terminates\ (ie(),b_{i+1},\ t)\ ^\wedge\ Terminates\ (ie(),b_{i+2},\ t)\ ^\wedge\ \ldots\ ^\wedge\ Terminates\ (ie(),b_j,\ t)\ \leftarrow Happens$$
$$(ie(),\ t)\ ^\wedge\ HoldsAt\ (Predominat(E) = emotion_k,\ t),\ \forall emotion_k \in E$$

Now, based on current beliefs and predominant emotions, virtual humans update their desires, initiating new ones and terminating the old ones, whenever an event, associated to an emotional change occurs. This is expressed as follows:

$$Initiates\ (e(),d_1,\ t)\ ^\wedge\ Initiates\ (e(),d_2,\ t)\ ^\wedge\ \ldots\ ^\wedge\ Initiates\ (e(),d_i,\ t)\ ^\wedge$$
$$Terminates\ (e(),d_i+1,\ t)\ ^\wedge\ Terminates\ (e(),d_{i+2},t)\ ^\wedge\ \ldots\ ^\wedge Terminates\ (e(),d_k,\ t)$$
$$\leftarrow Happens\ (e(),\ t)\ ^\wedge\ HoldsAt\ (b_1,\ t)\ ^\wedge\ HoldsAt\ (b_2,\ t)\ ^\wedge\ \ldots\ ^\wedge\ HoldsAt\ (b_j,\ t)\ ^\wedge$$
$$HoldsAt\ (Predominat(E) = emotion_k,\ t),\ \forall emotion_k \in E$$

Given current desires and after a deliberative process denoted by event *selectIntention* (), an agent commits a plan to attain a goal. Next is event calculus formalization:

$$Initiates\ (selectIntention(),\ decision,\ t)\ ^\wedge$$
$$Intention_k = \{Happens\ (action()_1,\ t_1)\ ^\wedge\ Happens\ (action()_2,\ t_2)\ ^\wedge\ \ldots\ ^\wedge$$
$$Happens\ (action()_m,\ t_n)\ \}\leftarrow \neg HoldsAt\ (decision,\ t)\ ^\wedge\ HoldsAt\ (d_j,\ t)\ ^\wedge$$
$$Happens\ (selectIntention(),\ t),\ \forall\ d_j \in D$$

As defined previously, plans can be interrupted. A plan interruption is managed by means of selecting a new intention according to the event that occurred. Whenever events occur, virtual humans reconsider their intentions. Finally, whenever a plan is satisfactorily achieved, that is to say, it arrives to perform the last action $action_m()$ of an intention, it releases the desire d_j that promotes the intention and enables the selection of a new plan to achieve a different goal.

$$Terminates\ (action()_m,\ decision,\ t_n)\ \wedge\ Terminates\ (action()_m,\ d_j,\ t_n) \leftarrow Happens\ (action()_m,\ t_n)$$

The previous event calculus formulas provide a framework to support an EBDI-based action selection process for autonomous virtual humans in dynamic virtual environments.

4.3 Case Study

The following example involves two virtual humans—Brian and Helena. Brian has a depressive personality and Helena has a paranoid personality. The interaction takes place in a discotheque, Helena is dancing alone, while Brian is walking to her, with the intention of flirting with her. When she realizes his intentions, she rejects him, gets angry and starts yelling at him. Then Brian starts crying, and finally she tries to console him. Initial states for both virtual humans are defined as follows:

$$Emotions_H = \{happiness\}$$
$$Initial\ Beliefs_H = \{I\ am\ Married,\ Nice\ Party\}$$
$$Initial\ Desires_H = \{Dance\ Alone,\ Be\ Aggressive,\ Be\ Polite\}$$
$$I=\{i_1=\{Happens\ (dance\ Favorite\ Song(),\ t_1)\ ^\wedge Happens\ (wait\ For\ Friends(),t_2)\},$$
$$i_2=\{Happens\ (yells\ At\ Aggressors(),\ t_1)^\wedge Happens\ (call\ My\ Friends(),\ t_2)^\wedge$$
$$Happens\ (go\ Home(),\ t_3)\},$$
$$i_3=\{Happens\ (ConsoleSad\ Person(),\ t_1)^\wedge Happens\ (Be\ Nice(),\ t_2)\ ^\wedge$$
$$Happens\ (say\ Bye\ Bye(),\ t_3)\}\}$$
$$Current\ Desire_H = Dance\ Alone$$

$$Emotions_B = \{happiness\}$$
$$Initial\ Beliefs_B = \{I\ am\ Handsome,\ I\ don't\ Have\ Girl\}$$
$$Initial\ Desires_B = \{Looking\ For\ Girls,\ Flirting\ With\ Girl,\ Kiss\ Girl\}$$
$$I=\{i_1=\{Happens\ (look\ For\ Girls(),\ t_1)\ ^\wedge Happens\ (walk\ At\ Her(),\ t_2)\},$$
$$i_2=\{Happens\ (kneel\ Down(),\ t_1)\ ^\wedge Happens\ (declare\ My\ Love(),\ t_2)\}$$
$$i_3=\{Happens\ (cry(),\ t_1)^\wedge Happens\ (wait\ Until\ Pity(),\ t_2)\ ^\wedge$$
$$Happens\ (begin\ Conversation(),\ t_3)\}\}$$
$$Current\ Desire_B = Looking\ For\ Girls$$

For this initial situation (Fig. 18a) both virtual humans have to be committed to a plan to attain their current goal (intention). This is done through the execution of an internal event called *select Intention()*.

For Helena:

$$Initiates\ (select\ Intention(),\ decision,\ t)\ ^\wedge$$
$$i_1=\{Happens\ (dance\ Favorite\ Song(),\ t_1)\ ^\wedge Happens\ (wait\ For\ Friends(),\ t_2)\}$$
$$\leftarrow \neg HoldsAt\ (decision,\ t)\ ^\wedge HoldsAt\ (Dance\ Alone,\ t)\ ^\wedge Happens\ (select\ Intention(),\ t)$$

For Brian:

$$Initiates\ (select\ Intention(),\ decision,\ t)\ ^\wedge$$
$$i_1=\{Happens\ (look\ For\ Girls(),\ t_1)\ ^\wedge Happens\ (walk\ At\ Her(),\ t_2)\}$$
$$\leftarrow \neg HoldsAt\ (decision,\ t)\ ^\wedge HoldsAt\ (Looking\ For\ Girls,\ t)\ ^\wedge Happens\ (select\ Intention(),\ t)$$

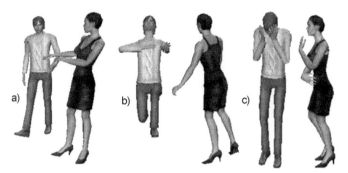

Fig. 18: Interaction between Brian and Helena: (a) Brian is walking to her, (b) Brian declares his love to her, (c) Helena rejects him and he begins to cry.

The current desire of Brian is to look for girls and the current desire of Helena is to dance alone. When Brian gets close to Helena, this is perceived as an external event for both virtual humans and it alters their beliefs.

In the case of Helena:

$$Initiates\ (walk\ At\ Her(),\ Aggressor\ In\ Front\ Of\ Me,\ t)\ \wedge$$
$$Terminates\ (walk\ At\ Her(),Nice\ Party,\ t) \leftarrow Happens\ (walk\ At\ Her(),\ t)$$

In the case of Brian:

$$Initiates\ (walk\ At\ Her(),\ She\ Is\ Pretty,\ t)\ \wedge$$
$$Terminates\ (walk\ At\ Her(),\ I\ don't\ Have\ Girl,\ t) \leftarrow Happens\ (walk\ At\ Her(),\ t)$$

After the update of beliefs, an update of desires takes places (Fig. 18b):
For Helena:

$$Initiates\ (walk\ At\ Her(),\ Be\ Aggressive,\ t)\ \wedge\ Terminates\ (walk\ At\ Her(),\ Dance\ Alone,\ t) \leftarrow$$
$$HoldsAt\ (Predominat\ (E) = happiness,\ t)\ ^\wedge HoldsAt\ (aggressor\ In\ Front\ Of\ Me,\ t)\ \wedge$$
$$HoldsAt\ (I\ am\ Married,\ t)\ \wedge\ Happens\ (walk\ At\ Her(),\ t)$$

For Brian:

$$Initiates\ (walk\ At\ Her(),\ Flirting\ With\ Girl,\ t)\ \wedge\ Terminates\ (walk\ At\ Her(),\ Looking\ Girls,\ t) \leftarrow$$
$$HoldsAt\ (Predominat(E) = happiness,\ t)\ ^\wedge HoldsAt\ (She\ Is\ Pretty,\ t)\ \wedge\ HoldsAt\ (I\ am$$
$$Handsome,\ t)\ \wedge\ Happens\ (walk\ At\ Her(),\ t)$$

With the update of desires comes the selection of intentions.

In the case of Helena:

$$Initiates\ (select\ Intention(),\ decision,\ t)\ ^\wedge i_2 = \{Happens\ (yells\ At\ Aggressors(),$$
$$t_1)^\wedge Happens\ (call\ My\ Friends(),\ t_2)\ \wedge\ Happens\ (go\ Home(),\ t_3)\} \leftarrow \neg HoldsAt\ (decision,\ t)$$
$$\wedge\ HoldsAt\ (Be\ Aggressive,\ t)\ \wedge\ Happens\ (select\ Intention(),\ t)$$

In the case of Brian:

$$Initiates\ (select\ Intention(),\ decision,\ t)\ \wedge\ i_2 = \{Happens\ (kneel\ Down(),\ t_1)\ \wedge\ Happens$$
$$(declare\ My\ Love(),\ t_2)\} \leftarrow \neg HoldsAt\ (decision,\ t)\ \wedge\ HoldsAt\ (Flirting\ With\ Girl,\ t)\ \wedge$$
$$Happens\ (select\ Intention(),\ t)$$

In this way, when new events are perceived by virtual humans, these produce the acquisition of new intentions (Fig. 18c). Thus, the course of action is followed until the virtual humans release their current desires.

A possible ending for this scenario could be that finally Brian kisses Helena:

$$\textit{Terminates (Brian Kiss Her(), decision, } t_n) \wedge \textit{Terminates (Brian Kiss Her(), Kiss Girl, } t_n)$$
$$\leftarrow \textit{Happens (Brian Kiss Her(), } t_n)$$

5. Conclusion

In recent years, the simulation of human behavior has become increasingly important as a field of interdisciplinary and multidisciplinary research of great interest. This field of research has attracted more and more research attention about realistic virtual humans in different applications of virtual reality and Artificial Intelligence based on autonomous and intelligent agents. These applications have recently focused on the creation of intelligent virtual humans, capable of expressing autonomous behavior and interacting realistically in 3D virtual environments based on the functioning of a perception-decision-action cycle. While there are a number of important aspects to simulate synthetic perception and decision making in virtual humans, this chapter focused on analyzing mechanisms to simulate characteristics and credible behaviors in autonomous virtual humans.

There exist many factors that influence the behavior of an individual, which make it impossible to predict with accuracy the actions that this person will expose and perform in the face of certain events. These events trigger emotional influences that change the affective state of an individual, giving the person the ability to generate an emotional response to the different situations experienced in the real world. This chapter discusses the various aspects of designing and modeling autonomous virtual humans and the most important behavioral architectures used to simulate human behavior. With the many advances in virtual reality and Artificial Intelligence technologies in recent years, decision making is becoming more important to model the mind of the virtual human as well as analyzing the captured data from the real human and modeling the interaction between virtual and real. Finally, a new model to simulate the human behavior in virtual humans was proposed. This model can be used in applications where it is necessary that virtual humans behave in a similar way as human beings do, according to their affective state, beliefs, desires, and intentions and the nature of the events they perceive from their environment.

References

Ali, N.S. and Nasser, M. (2017). Review of virtual reality trends (previous, current, and future directions), and their applications, technologies and technical issues. ARPN Journal of Engineering and Applied Sciences, 12(3): 783–789.

Amunts, K., Ebell, C., Muller, J., Telefont, M., Knoll, A. and Lippert, T. (2016). The human brain project: Creating a European research infrastructure to decode the human brain. Neuron, 92(3): 574–581.

Anthes, C., García-Hernández, R.J., Wiedemann, M. and Kranzlmüller, D. (2016). State of the art of virtual reality technology. IEEE Aerospace Conference, pp. 1–19, IEEE.

Armstrong, E. and Falk, D. (2012). Primate Brain Evolution: Methods and Concepts. Springer Science & Business Media.

Asensio, J.M., Peralta, J., Arrabales, R., Bedia, M.G., Cortez, P. and Peña, A.L. (2014). Artificial intelligence approaches for the generation and assessment of believable human-like behaviour in virtual characters. Expert Systems with Applications, 41(16): 7281–7290.

Bajcsy, R., Aloimonos, Y. and Tsotsos, J.K. (2018). Revisiting active perception. Autonomous Robots, 42(2): 177–196.

Balint, T. and Allbeck, J.M. (2013). What's going on? Multi-sense attention for virtual agents. pp. 349–357. *In*: International Workshop on Intelligent Virtual Agents. Berlin, Germany: Springer.

Ball, G. and Breese, J. (2000). Emotion and personality in a conversational agent. pp. 189–219. *In*: Cassell, J., Sullivan, J., Prevost, S. and Churchill, E. (eds.). Embodied Conversational Agents. MIT Press.

Blauert, J. (1997). Spatial Hearing: The Psychophysics of Human Sound Localization. MIT Press.

Bogdanovych, A. and Trescak, T. (2016). Generating needs, goals and plans for virtual agents in social simulations. International Conference on Intelligent Virtual Agents, pp. 397–401, Los Angeles, USA: Springer, Cham.

Bourgais, M., Taillandier, P. and Vercouter, L. (2016). An agent architecture coupling cognition and emotions for simulation of complex systems. Social Simulation Conference, pp. 1–12, Rome, Italy: HAL.

Burden, D. and Savin-Baden, M. (2019). Virtual Humans: Today and Tomorrow. CRC Press.

Caillou, P., Gaudou, B., Grignard, A., Truong, C.Q. and Taillandier, P. (2015). A simple-to-use BDI architecture for agent-based modeling and simulation. Advances in Social Simulation, pp. 15–28, Springer.

Conde, T., Tambellini, W. and Thalmann, D. (2003). Behavioral animation of autonomous virtual agents helped by reinforcement learning. International Workshop on Intelligent Virtual Agents, pp. 175–180. Kloster Irsee, Germany: Springer.

Conde, T. and Thalmann, D. (2006). Learnable behavioural model for autonomous virtual agents: Low-level learning. Fifth International Joint Conference on Autonomous Agents and Multiagent Systems, pp. 89–96. Hakodate, Japan: ACM.

Costa, P.T. and McCrae, R.R. (1992). Normal personality assessment in clinical practice: The NEO personality inventory. Psychological Assessment, 4(1): 5–13.

DeVault, D., Artstein, R., Benn, G., Dey, T., Fast, E., Gainer, A. and Lucas, G. (2014). SimSensei Kiosk: A virtual human interviewer for healthcare decision support. International Conference on Autonomous Agents and Multi-agent Systems, pp. 1061–1068, Paris, France: International Foundation for Autonomous Agents and Multiagent Systems.

Egges, A., Kshirsagar, S. and Magnenat-Thalmann, N. (2004). Generic personality and emotion simulation for conversational agents. Computer Animation and Virtual Worlds, 15(1): 1–39.

Ekman, P.E. and Davidson, R.J. (1994). The Nature of Emotion: Fundamental Questions. Oxford University Press.

El Saddik, A., Orozco, M., Eid, M. and Cha, J. (2011). Haptics Technologies: Bringing touch to Multimedia, Ottawa, Canada: Springer Science & Business Media.

Feng, A.W., Xu, Y. and Shapiro, A. (2012). An example-based motion synthesis technique for locomotion and object manipulation. ACM SIGGRAPH Symposium on Interactive 3D Graphics and Games, pp. 95–102, Costa Mesa, California: ACM.

García-Rojas, A., Gutiérrez, M. and Thalmann, D. (2008). Simulation of individual spontaneous reactive behavior. 7th International Joint Conference on Autonomous Agents and Multiagent Systems, 1: 143–150, Estoril, Portugal: International Foundation for Autonomous Agents.

Gillies, M.F. (2001). Practical Behavioral Animation Based on Vision and Attention (Doctoral Dissertation), University of Cambridge.

Gratch, J., Hartholt, A., Dehghani, M. and Marsella, S. (2013). Virtual humans: A new toolkit for cognitive science research. Proceedings of the Annual Meeting of the Cognitive Science Society, 35(35): 41–42.

Grillon, H. and Thalmann, D. (2009). Simulating gaze attention behaviors for crowds. Computer Animation and Virtual Worlds, 20(2-3): 111–119.

Gu, E. and Badler, N.I. (2006). Visual attention and eye gaze during multiparty conversations with distractions. International Workshop on Intelligent Virtual Agents, pp. 193–204. Marina Del Rey, CA, USA: Springer.

Gutierrez-Garcia, J.O. and López-Neri, E. (2015). Cognitive computing: A brief survey and open research challenges. 3rd International Conference on Applied Computing and Information Technology/2nd International Conference on Computational Science and Intelligence, pp. 328–333, IEEE.

Handel, O. (2016). Modeling dynamic decision-making of virtual humans. Systems, 4(1): 1–26.

Harland, J., Morley, D.N., Thangarajah, J. and Yorke-Smith, N. (2017). Aborting, suspending, and resuming goals and plans in BDI agents. Autonomous Agents and Multi-Agent Systems, 31(2): 288–331.

Haubrich, T., Seele, S., Herpers, R., Bauckhage, C. and Becker, P. (2014). Synthetic perception for intelligent virtual agents. First ACM SIGCHI Annual Symposium on Computer-human Interaction in Play, pp. 421–422, Toronto, Ontario, Canada: ACM.

Herrero, P. and de Antonio, A. (2003). Introducing human-like hearing perception in intelligent virtual agents. Second International Joint Conference on Autonomous Agents and Multiagent Systems, pp. 733–740, Melbourne, Australia: ACM.

Herrero, P. and de Antonio, A. (2004). Modeling intelligent virtual agent skills with human-like senses. International Conference on Computational Science, pp. 575–582, Berlin, Germany: Springer.

Hill, R. (1999). Modeling perceptual attention in virtual humans. 8th Conference on Computer Generated Forces and Behavioral Representation, pp. 563–573, Orlando, FL.

Iglesias, A. and Luengo, F. (2007). AI framework for decision modeling in behavioral animation of virtual avatars. 7th International Conference on Computational Science, pp. 89–96, Beijing, China: Springer.

Itti, L., Dhavale, N. and Pighin, F. (2003). Realistic avatar eye and head animation using a neurobiological model of visual attention. Applications and Science of Neural Networks, Fuzzy Systems, and Evolutionary Computation VI, pp. 64–79, San Diego, California, United States: SPIE.

Ivanović, M., Budimac, Z., Radovanović, M., Kurbalija, V., Dai, W., Bădică, C. and Mitrović, D. (2015). Emotional agents—state-of-the-art and applications. Computer Science and Information Systems, 12(4): 1121–1148.

Jerald, J. (2015). The VR Book: Human-centered design for Virtual Reality. Morgan & Claypool.

Jiang, H., Vidal, J.M. and Huhns, M.N. (2007). EBDI: An architecture for emotional agents. 6th International Joint Conference on Autonomous Agents and Multiagent systems, pp. 1–19, Honolulu, Hawaii: ACM.

Jiang, H. (2008). From Rational to Emotional Agents: A Way to Design Emotional Agents. Saarbrücken, Germany: VDM Verlag.

Johnson, W.L. and Rickel, J. (1997). Steve: An animated pedagogical agent for procedural training in virtual environments. ACM SIGART Bulletin, 8(1-4): 16–21, New York, NY, USA: ACM.

Khullar, S.C. and Badler, N.I. (2001). Where to look? Automating attending behaviors of virtual human characters. Autonomous Agents and Multi-Agent Systems, 4(1): 9–23.

Kim, Y., Hill, R.W. and Traum, D.R. (2005). A computational model of dynamic perceptual attention for virtual humans. 14th Conference on Behavior Representation in Modeling and Simulation, pp. 1–8, Universal City, CA: University of Southern California.

Kotseruba, I., Gonzalez, O.J. and Tsotsos, J.K. (2016). A Review of 40 Years of Cognitive Architecture Research: Focus on Perception, Attention, Learning and Applications.

Kshirsagar, S. and Magnenat-Thalmann, N. (2002). A multilayer personality model. 2nd International Symposium on Smart Graphics, pp. 107–115, ACM.

Kuffner, J.J. and Latombe, J.C. (1999a). Perception-based Navigation for Animated Characters in Real-Time Virtual Environments, 1–22.

Kuffner, J.J. and Latombe, J.C. (1999b). Fast synthetic vision, memory, and learning models for virtual humans. Computer Animation, pp. 118–127, Geneva, Switzerland: IEEE.

Kuiper, D.M. and Wenkstern, R.Z. (2013). Virtual agent perception combination in multi-agent based systems. International Conference on Autonomous Agents and Multi-agent Systems, pp. 611–618, St. Paul, MN, USA: ACM.

Lee, J., Won, J. and Lee, J. (2018). Crowd simulation by deep reinforcement learning. 11th Annual International Conference on Motion, Interaction, and Games, pp. 1–7, Limassol, Cyprus: ACM.

Lieto, A., Bhatt, M., Oltramari, A. and Vernon, D. (2018). The role of cognitive architectures in general artificial intelligence. Cognitive Systems Research, 1–3.

Liu, Z. and Lu, Y.S. (2008). A motivation model for virtual characters. 2008 International Conference on Machine Learning and Cybernetics, 5: 2712–2717, Kunming, China: IEEE.

Magnenat-Thalmann, N. and Thalmann, D. (2005). Handbook of Virtual Humans. John Wiley & Sons.

Mayer, J.D., Caruso, D.R. and Salovey, P. (2016). The ability model of emotional intelligence: Principles and updates. Emotion Review, 8(4): 290–300.

Meneguzzi, F. and De Silva, L. (2015). Planning in BDI agents: A survey of the integration of planning algorithms and agent reasoning. The Knowledge Engineering Review, 30(1): 1–44.

Minsky, M. (2006). The Emotion Machine: Commonsense Thinking, Artificial Intelligence, and the Future of the Human Mind. New York, New York, USA: Simon and Schuster.

Müller, V.C. and Bostrom, N. (2016). Future progress in artificial intelligence: A Survey of expert opinion. pp. 553–571. *In*: Müller, V.C. (ed.). Fundamental Issues of Artificial Intelligence. Berlin, Germany: Springer.

Newell, A. (1994). Unified Theories of Cognition. Harvard University Press.

Noser, H. and Thalmann, D. (1995). Synthetic vision and audition for digital actors. Computer Graphics Forum, 14(3): 325–336.

Nunnari, F. and Heloir, A. (2017). Generation of virtual characters from personality traits. 17th International Conference on Intelligent Virtual Agents, pp. 301–314, Stockholm, Sweden: Springer.

Orozco, H., Ramos, F., Fernández, V., Gutiérrez, O., Ramos, M.A. and Thalmann, D. (2010). A cognitive model for human behavior simulation in EBDI virtual humans. Second International Conference on Agents and Artificial Intelligence, pp. 104–111, Valencia, Spain: Springer.

Orozco, H., Ramos, F., Ramos, M. and Thalmann, D. (2011). An action selection process to simulate the human behavior in virtual humans with real personality. The Visual Computer, 27(4): 275–285.

Ortony, A., Clore, G.L. and Collins, A. (1998). The Cognitive Structure of Emotions. Cambridge, UK: Cambridge University Press.

Pan, X. and Hamilton, A.F. (2018). Why and how to use virtual reality to study human social interaction: The challenges of exploring a new research landscape. British Journal of Psychology, 109(3): 395–417.

Patkos, T., Plexousakis, D., Chibani, A. and Amirat, Y. (2016). An event calculus production rule system for reasoning in dynamic and uncertain domains. Theory and Practice of Logic Programming, 16(3): 325–352.

Peters, C. and O'Sullivan, C. (2002). Synthetic vision and memory for autonomous virtual humans. Computer Graphics Forum, 21(4): 743–752.

Peters, C. and O'Sullivan, C. (2003). Bottom-up visual attention for virtual human animation. 11th IEEE International Workshop on Program Comprehension, pp. 111–117, New Brunswick, NJ, USA, USA: IEEE.

Peters, C., Pelachaud, C., Bevacqua, E., Mancini, M. and Poggi, I. (2005). A model of attention and interest using gaze behavior. International Workshop on Intelligent Virtual Agents, pp. 229–240, Kos, Greece: Springer.

Pudane, M., Lavendelis, E. and Radin, M.A. (2017). Human emotional behavior simulation in intelligent agents: Processes and architecture. Procedia Computer Science, 104: 517–524.

Rabie, T.F. (2002). Autonomous perception systems for dynamic virtual environments. IEEE International Workshop HAVE Haptic Virtual Environments, pp. 85–90, Ottawa, Ontario, Canada: IEEE.

Renault, O., Thalmann, N.M. and Thalmann, D. (1990). A vision-based approach to behavioural animation. The Journal of Visualization and Computer Animation, 1(1): 18–21.

Rosenfeld, A. and Kraus, S. (2018). Predicting Human Decision-Making: From Prediction to Action. Morgan & Claypool Publishers.

Ruhland, K., Andrist, S., Badler, J., Peters, C., Badler, N., Gleicher, M. and Mcdonnell, R. (2014). Look me in the eyes: A survey of eye and gaze animation for virtual agents and artificial systems. Eurographics State-of-the-Art Report, pp. 69–91, Strasbourg, France: HAL.

Saberi, M., Bernardet, U. and DiPaola, S. (2014). An architecture for personality-based, nonverbal behavior in affective virtual humanoid character. Procedia Computer Science, 41: 204–211.

Sellbom, M. (2019). The MMPI-2-restructured form (MMPI-2-RF): Assessment of personality and psychopathology in the twenty-first century. Annual Review of Clinical Psychology, 15: 149–177.

Steel, T., Kuiper, D. and Wenkstern, R.Z. (2010). Virtual agent perception in multi-agent-based simulation systems. IEEE/WIC/ACM International Conference on Web Intelligence and Intelligent Agent Technology, pp. 453–456, Toronto, Ontario, Canada: IEEE.

Steunebrink, B.R., Dastani, M. and Meyer, J.J. (2009). The OCC model revisited, Workshop on emotion and computing. Association for the Advancement of Artificial Intelligence, pp. 1–8, Paderborn, Germany.

Su, W.P., Pham, B. and Wardhani, A. (2007). Personality and emotion-based high-level control of affective story characters. IEEE Transactions on Visualization and Computer Graphics, 13(2): 281–293.

Swartout, W.R., Gratch, J., Hill Jr, R.W., Hovy, E., Marsella, S., Rickel, J. and Traum, D. (2006). Toward virtual humans. AI Magazine, 27(2): 96–108.

Swartout, W.R. (2016). Virtual humans as centaurs: Melding real and virtual. International Conference on Virtual, Augmented and Mixed Reality, pp. 356–359, Toronto, Canada: Springer.

Thalmann, D., Noser, H. and Huang, Z. (1997). Autonomous virtual actors based on virtual sensors. Creating Personalities for Synthetic Actors, pp. 25–42, Springer.

Vasant, P. and DeMarco, A. (2015). Handbook of research on artificial intelligence techniques and algorithms. Information Science Reference.

Vosinakis, S. and Panayiotopoulos, T. (2005). A tool for constructing 3D environments with virtual agents. Multimedia Tools and Applications, 25(2): 253–279.

Wu, J. and Kraemer, P. (2016). Virtual psychology: An overview of theory, research, and future possibilities. pp. 293–307. *In*: Sivan, Y. (ed.). Handbook on 3D3C Platforms—Applications and Tools for Three-dimensional Systems for Community, Creation and Commerce. Springer, Cham.

Yu, Q. and Terzopoulos, D. (2007). A decision network framework for the behavioral animation of virtual humans. 2007 ACM SIGGRAPH/Eurographics Symposium on Computer Animation, pp. 119–128, San Diego, California: Eurographics Association.

Yumak, Z., Ren, J., Thalmann, N.M. and Yuan, J. (2014). Modeling multi-party interactions among virtual characters, robots, and humans. Presence: Teleoperators and Virtual Environments, 23(2): 172–190.

Yumak, Z., van den Brink, B. and Egges, A. (2017). Autonomous social gaze model for an interactive virtual character in real-life settings. Computer Animation and Virtual Worlds, 28(3–4).

Chapter **5**

An Internal Model for Characters in Virtual Environments
Emotions, Moods and Personality

Frank Julca, Gonzalo Méndez* and Raquel Hervás

1. Introduction

Emotions influence the interactions between people because they are present on everything we experience daily. They affect our mental state, facial and body expressions, decision making, and social interactions. During the last few years, there has been a significant increase in research on computational models of emotional processes, stemming from the fields of emotion research in psychology, cognitive science, philosophy, artificial intelligence, and computational science. This interest has been driven both by their potential for basic research on human emotions and cognition and by the perspective of an ever-growing range of applications. One of such applications is the realistic modeling of characters in virtual environments and which are subsequently able to exhibit a varied range of non-monotonous, human-like behaviors.

The interaction with computers should be as natural as possible, and we, as users, tend to expect that computers to react and behave as human beings would. So it is of utmost importance that computer applications are able to interpret and simulate emotions in a natural way. One of the tools provided by computational science, that allows us to model the approximate reasoning mechanisms that humans use, is fuzzy logic. However, few research works have been carried out on the use of fuzzy logic to model and simulate emotions. The most important models turn out to be too theoretical and fail to put forward a way to combine emotions with moods and personality traits to obtain an even more complex and realistic emotional simulation.

The internal model of emotions for virtual characters proposed in this chapter takes as its main reference the theoretical basis of fuzzy logic applied to emotions of existing computational models. The combination with mood is the most advanced step that these

Universidad Complutense de Madrid, Spain.
* Corresponding author: gmendez@fdi.ucm.es

models have developed. The main objective of the work described in this chapter is to simulate emotions as a reaction to events that occur in the environment, enhancing the behavior of characters through the additional use of both mood and personality traits.

To achieve these goals, overviews of the main psychological theories of emotions are provided in Section 2. Subsequently, an overview of the state-of-the-art in computational models of emotion is presented in Section 3, which examines the most relevant and influential models in the field. These two sections do not intend to be a comprehensive study of the field for which, the reader can refer to (Marsella et al., 2010). In addition, in Section 4, the foundations of fuzzy logic are shown and how it is applied to the generation of emotions of virtual characters. In Section 5, we present our model for virtual characters, providing details on how emotions, moods and personality have been modeled and integrated. After that, Section 6 provides details of our implementation of this model, which empirically validates it and which serves as a basis for establishing the experimental values of some of the constants used in the model. Finally, some conclusions obtained from our work are presented.

2. Psychological Models of Emotions, Moods and Personality

In this section we provide an overview of the psychological basis that grounds the computational models presented in the following section for emotions, moods and personality, describing their purpose and classifications.

2.1 Emotions

Although classic philosophers, like Plato or Descartes, considered that emotions were not part of human intelligence but an obstacle in human thinking, modern psychologists acknowledge emotions as a positive component of human cognition (Ekman, 1992). Modern research defends that emotions have a great impact on cognition, memory and beliefs (Forgas, 1995), to the point of considering that not having emotional response capabilities implies that it is more difficult to make good decisions (Damasio and Sutherland, 2008).

The main goal of a psychological theory of emotions is to describe the processes related to the creation of emotions within an individual, even if these processes occur in a social context and not within the individual (Frijda et al., 2000). Despite extensive research on emotions from a psychological point of view, it is not possible to establish a theory of emotions that is considered 'correct' and, therefore, widely accepted by the scientific community (Reisenzein et al., 2013). For example, Kenneth Strongman, in his more than fourteen years of experience in the field of psychology, concludes that there are more than 150 different emotion theories (Strongman, 2003). In the same way, theories can be described in different ways and, although they may not seem similar, they do not necessarily contradict each other.

It is usually considered that there are four major trends in the psychology of emotions, corresponding to a classification of the most famous and influential theories used to model emotions: somatic, discrete, dimensional, and cognitive. Somatic theories consider emotions as experiences similar to sensations; discrete theories suppose that emotions are something innate in all human beings and therefore have their own processing mechanisms hard-coded in the human brain; dimensional theories argue that emotions are mainly psychological labels that are assigned to mental and physical states, and identify them with a position in a two-dimensional or tri-dimensional space. Finally, cognitive theories assume that there are physiological, subjective and cognitive components involved in the simulation of emotions, and that mental representations are required to explain the importance of cognition in emotions.

Appraisal theory (Marsella et al., 2010) is a cognitive theory that is one of the most prominent theories used for emotion modelling. Appraisal theory states that emotions arise from the interaction between things we consider important and events that happen in our environment, so they cannot be easily explained by focusing only on the individual or his environment. Therefore, it is not only our relationship with the environment or the relative importance of things, but also the interaction between both factors that produces our emotional responses. Within the field of appraisal theory, there are even more specific theories. The most important ones, and the basis of many computational models of emotions, are the theories of Frijda (Frijda, 1986), OCC (Ortony et al., 1990), and Lazarus (Lazarus, 1991).

Frijda's theory was defined in 1986, when Nico H. Frijda wrote his famous book, *The Emotions'* (Frijda, 1986), where he describes the development of his psychological theory of emotions. His book had two main goals: one was to find a definition of emotions that could be widely accepted, and the other was to understand the origin and function of emotions, as well as the conditions that are necessary for them to appear.

A short time later, Ortony et al. (Ortony et al., 1990) developed a joint theory (usually called OCC model) that is currently widely used to build computational models of emotions (Hudlicka, 2011). Their main objective was to describe the cognitive architecture of emotions so that they could be easily defined, and then establish relationships between them. In that way, it should be possible to develop a computational model of emotions that could be used in Artificial Intelligence.

Later, in 1991, Richard Lazarus finished his research on stress (Lazarus, 1991), which was a very relevant psychological problem during World War II and that was when Lazarus began his investigations. This led him to the study of emotions and the subsequent development of coping—a key concept in his research that is usually related to stress.

Although all these theories try to define what emotions are, the OCC model provides a clearer definition: emotions are reactions to events, agents, or objects and their origin is determined by the way in which the situation that triggers them is interpreted by the individual (Ortony et al., 1990).

2.2 *Moods*

Moods can be considered as emotional situations occurring at a particular time and whose duration is longer than that of emotions (it can last for hours or days). In the field of emotion modelling, moods can act as a filter in the process of modelling emotions, so that the generated emotional responses are as realistic as possible. Traditionally, the mood has been modeled as a part of the emotional state, that is, as if it just were an additional emotion. However, moods differ from emotions because they reflect a more stable and lasting emotional state of an individual, and therefore must be treated separately. From the point of view of the dimensional theory of Mehrabian (Mehrabian, 1995), a mood can be defined as an average of the emotional state of a person in a variety of daily life situations. This theory is also one of the most popular recent computational models of emotions due to the simplification of mood in three traits: pleasure, arousal and dominance. These three characteristics form a three-dimensional space for mood in which each of the dimensions is measured in numerical values, ranging from −1.0 to 1.0. In this way, different combinations of these features can be established to classify different mood states. Examples can be seen in Table 1, in which +P and −P are used for pleasant and unpleasant, +A and −A for exciting and gentle, and +D and −D for dominant and submissive.

Each type of mood can then be defined by a value in each of the dimensions of pleasure, arousal and dominance. For example, a person is in a relaxed mood if the values of pleasure and dominance are positive, and that of arousal is negative.

Table 1: Moods according to the PAD dimensional space.

+P+A+D Exuberant	−P−A−D Boring
+P+A−D Dependent	−P−A+D Disdainful
+P−A+D Relaxed	−P+A−D Anxious
+P−A−D Submissive	−P+A+D Boring

2.3 Personality

Personality in virtual agents can be expressed in many ways: through language (Mairesse and Walker, 2010) or non-verbal behavior, such as facial expressions (Bee et al., 2010) or gestures (McRorie et al., 2012). Personality theories have always tried to specify how many personality traits human beings have. Gordon Allport proposed a list of more than 4,000 features (Allport, 1937), Raymond Cattell considered 16 personality factors (Cattell, 1957), and Hans Eysenck summarized them into a theory with only three factors (Eysenck, 1991). Many researchers consider that Cattell's theory was too complicated and that Eysenck's had too limited a scope. As a result of these investigations arose the theory that is known in psychology as the Big Five model.

The Big Five (Barrick and Mount, 1991) is a theory that examines the structure of personality by considering five factors or personal traits represented like five different dimensions: extraversion, agreeableness, open-ness to experience, conscientiousness and neuroticism. Each of the five personality dimensions represents a range between two extremes. For example, extraversion represents the continuous range between the most extroverted and the most introverted trait. The characteristics of each of the dimensions are briefly described below:

- **Openness to experience:** People who have this trait tend to be curious about the world and other people, and are eager to learn new things and enjoy new experiences. On the other hand, people who are not very open are much more traditional and resist new ideas because they do not like changes.

- **Conscientiousness:** High conscientious people have control over their impulses, plan their tasks with time and think about how their behavior can affect others. On the opposite side, low conscientious people overlook important tasks and, in general, do not complete things they are incharge of.

- **Extraversion:** People, who have a high level of extraversion, tend to be more sociable, enjoy meeting new people and initiating new conversations. Introverts are reserved and find it difficult to start a conversation. They prefer loneliness because social events can be exhausting for them.

- **Agreeableness:** People who have a high level of agreeableness tend to be more cooperative and feel empathy and concern for others, while those who have a low level tend to be more competitive and may come to be argumentative or untrustworthy.

- **Neuroticism:** People with a high level of neuroticism tend to get irritated and depressed more easily, with a tendency of experiencing drastic changes in mood. On the contrary, a person who is not neurotic is emotionally stable and rarely feels sad or depressed.

The Big Fives schema is used in many modern computational models that consider personality traits for the simulation of emotions. For example, ALMA (Gebhard, 2005) includes this model in the personality profile of its system and relates personality to emotions and moods through a mapping of values that fit the three-dimensional PAD model on which its architecture is based.

3. Computational Models of Emotions, Moods and Personality

A computational model of cognition is responsible for providing the necessary information to understand cognitive functions, describing the entire process through algorithms and programs (Sun, 2008). In this work, we are interested in computational models of emotions, and related issues, like moods and personality.

Computational models of emotions are usually based on psychological theories as a basis for building an effective computational model. They usually try to define what an emotion consists of, or how to understand emotional states, like fear or happiness. However, as we have seen in Section 2.1, there are many psychological theories about emotions, which differ in points as important as the involved cognitive processes or the representations of emotions. That makes the development of a computational model of emotions a quite complex task.

In the field of Artificial Intelligence, computational models of emotions are used to design autonomous agents that are able to interact with other agents or people, using verbal and non-verbal behavior (Marsella and Gratch, 2014). Therefore, they must be able to address how emotions arise and change in relation to different conditions—from simple events to complex situations that involve more agents. In addition, emotions must be directly related, both to the dynamics of how the environment works and to the mechanics of the individual's cognitive and behavioral processes.

3.1 Most Influential Computational Models and Architectures

Since there are many computational models of emotions used in affective computing, it is almost impossible to compare them all. For this reason, in this work, this comparison has been limited to those models that have had a high influence in this area, analyzing the general properties of each of them. These models are:

- **FLAME (Fuzzy Logic Adaptive Model of Emotions)** (El-Nasretal, 2000): FLAME is a model of emotions and moods that use fuzzy logic to process events and obtain new emotional states. This model is based on the OCC model and includes algorithms to learn patterns of events, expectations and relationships between objects, among others. FLAME is widely used to control the behavior of agents in virtual environments.

- **EMA (Emotions and Adaptation)** (Marsella and Gratch, 2009): EMA is a computational model of emotions, designed to provide very realistic emotional responses in virtual agents. This model is widely influenced by Lazarus' theories.

- **FAtiMA (Fear not AffectTIve Mind Architecture)** (Diasetal, 2014): FAtiMA is a cognitive architecture based on the OCC model that is specially designed to create intelligent agents that are able to use motions and personality in order to influence their behavior and be used in different scenarios. Due to its complexity, in 2014, a modular and simpler version of FAtiMA was developed that only contains some of its components to make the architecture more understandable and usable.

- **ALMA (A Layered Model of Affect)** (Gebhard, 2005): ALMA is a programming framework that implements the OCC model and the Mehrabian dimensional theory (PAD) and incorporates emotions, moods, and personality. This tool allows developers to build their own computer models of emotions for multiple applications.

It is important to note that each of these models has different objectives. Some are specially designed to provide a formal basis for a theory; others aim to develop a programming tool, such as ALMA; and others, such as EMA, seek to develop an application. In addition, we can see that the OCC model is the most popular one. In this work we have used FLAME as the computational model of emotions for our virtual agents. Among other features, the use of fuzzy logic makes it very useful when treating with uncertainty issues related to emotional states.

3.2 FLAME

FLAME (Fuzzy Logic Adaptive Model of Emotions) (El-Nasr et al., 2000) includes two new features with respect to the OCC model in which it is based (Roseman et al., 1990; Ortony et al., 1990). The first one is the representation of emotions with fuzzy logic in order to perform a mapping of events and expectations on behaviors and emotional states, using rules. The second is the incorporation of machine learning techniques, so the agents relate to the objects and events of the environment as well as to the user's objectives and expectations, giving rise to different responses.

Regarding the architecture of the model, FLAME defines three main components: an emotional component, a learning component and a decision-making component. As per the representation of the architecture in Fig. 1, the process starts when the agent perceives external events from the environment. These perceptions are processed by the learning and emotional components. However, the emotional component uses expectations and relationships between events and objectives to generate emotions. In addition, an emotional behavior is generated, which will be used as an input in the decision-making component. The final action to be performed by the agent will have to consider the mood, emotional state and emotional behavior obtained during this process.

In Fig. 2 the processes that make up the emotional component are shown. When an event happens, the agent obtains information from the environment. This information is propagated through the component, so that the output produced by each process serves as an input to the next one. First, the evaluation process is responsible for identifying the objectives that are affected by the event and for associating the degree of impact that

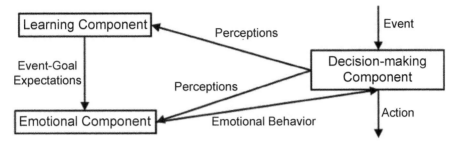

Fig. 1: Abstract representation of an agent architecture (El-Nasr et al., 2000).

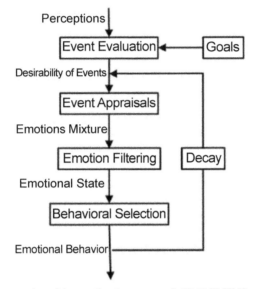

Fig. 2: Representation of the emotional component in FLAME (El-Nasr et al., 2000).

the event has on these objectives. Next, the level of desirability of the event is computed according to the impact and the importance of the affected objectives. All this process is performed using fuzzy logic rules.

Once the level of desirability is calculated, it is transferred to the classification process to calculate the new emotional state of the agent. Subsequently, the filtering component is responsible for considering the mood of the agent to detect dependencies with previous emotions. Finally, the reaction of the agent to the event is selected from the emotional state obtained in previous processes. The emotional state and intensity of the emotions of the agent will not be immutable until a new event happens, but will decrease as time goes by.

4. Fuzzy Logic

When describing the emotional state of a person, it is quite usual to use vague expressions, such as Peter is very tired or Lucy is a bit angry. A human being is able to understand

and interpret these undetermined expressions without problems. However, a machine will have trouble to interpret information that is not expressed by using absolute values. Fuzzy logic is very useful in this kind of contexts, due to its ability to solve complex and undefined problems, as opposed to the difficulty of solving them, using the traditional procedures of classical logic.

4.1 Fuzzy Sets

The term 'fuzzy set' was first used by Lotfi A. Zadeh in his theory of fuzzy sets (Klir and Yuan, 1995). This theory was presented as a generalization of the classical set theory and its main function is to associate the degree of membership of elements in a set. In this way, a proposition is not totally false or true, but it can be partially satisfied. This association process is known as fuzzification.

Fuzzy sets can be seen as a generalization of normal or crisp sets. Fuzzy logic is a multivalued logic, so it is necessary to use a range to define the continuous values that an element can have. Therefore, it is possible to use this range of values to indicate the degree of membership of an element in the set. Therefore, the membership function μA from a fuzzy set A has the following definition:

$$\mu A = X \rightarrow [0, 1]$$

where $\mu A(x) = 1$ indicating that x completely belongs to A, $\mu A(x) = 0$ is the opposite case where x is not in A and $0 < \mu A(x) < 1$ if x is partially in set A. The degree of membership of element x to the fuzzy set A is calculated in this way.

In addition, the membership functions allow graphical representation of fuzzy sets. When the calculations are not very complex, it is usual to choose simple functions to facilitate the operations of the system. The most used functions are the triangular and trapezoidal ones, as they have parameters that can be adjusted to achieve the desired domain. The triangular functions are defined in the following way:

$$\mu A(x) = \begin{cases} 0, & x \leq a \\ \dfrac{x-a}{m-a}, & a < x \leq m \\ \dfrac{b-x}{b-m}, & m < x < b \\ 0, & x \geq b \end{cases}$$

with a and b being the lower and upper limits, respectively. The medium value m is the maximum value of the function, which can be asymmetric (Fig. 3a). The trapezoidal function adds extra parameters to its definition to complete a trapezoidal form:

$$\mu A(x) = \begin{cases} 0, & (x < a) \text{ or } (x > d) \\ \dfrac{x-a}{b-a}, & a \leq x \leq b \\ 1, & b \leq x \leq c \\ \dfrac{d-x}{d-c}, & c \leq x \leq d \end{cases}$$

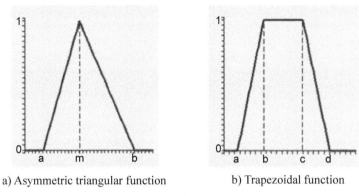

a) Asymmetric triangular function   b) Trapezoidal function

Fig. 3: Representation of membership functions.

with a and d being the lower and upper limit, respectively, and b and c the maximum values of lower and upper limits (Fig. 3b). The triangular function can be seen as a particular case of the trapezoidal function, where b and c values are equivalent.

4.2 Use of Linguistic Variables in Fuzzy Rules

Linguistic variables are variables that represent elements without a clear definition through words or sentences in natural language. Mainly, they are composed of a primary term and a modifier. Primary terms (high, strong, sad, etc.) are used to construct fuzzy sets, and modifiers (very, too, a little, etc.) are used to calculate the other fuzzy sets of compound terms. For example, in Fig. 4 we can see the linguistic variable height with fuzzy sets—very low, low, medium, high and very high to represent the values the variable can take.

In order to represent natural language sentences in terms of linguistic variables, we use fuzzy propositions. This concept is based on replacing the possible variables and their corresponding values with symbols that help to process the propositions. For example, the fact Peter is not a millionaire can be represented as P is not M, using the logical connector 'not'. This transformation facilitates the processing of facts through the so-called fuzzy rules expressed in the form of an if-then sentence, like IF <fuzzy proposition> THEN <fuzzy proposition>. In classic rules, if the antecedent is fulfilled, then the consequent will also be fulfilled. On the contrary, fuzzy propositions are valid, depending on the degree of membership; for example, in Fig. 4 a height of 185 cm would have a degree of membership of 0.5 in the fuzzy set as very high.

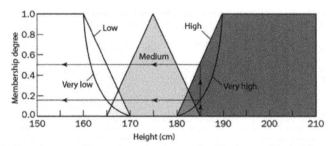

Fig. 4: Use of very modifier to determine the membership degree of the height variable.

One of the most influential inference types is the inference of Mamdani (Pourjavad and Mayorga, 2017), used in many computational models of emotions, such as FLAME. Given that the system described in this chapter is based on this model, it is convenient to explain in more detail the four steps followed through Mamdani's inference:

1. *Fuzzification of Input Variables*: The crisp values of input variables are checked to calculate the degree of membership they have in each of the previously defined fuzzy sets.

2. *Evaluation of Fuzzy Rules*: Once the fuzzy input variables are obtained, all the rules are evaluated by checking if the antecedents are fulfilled. The final value obtained will be used to determine the fuzzy set to which the consequent belongs.

3. *Obtaining Output Variables from the Rules*: Membership degrees from all the consequents must be combined, so that a single fuzzy set is obtained for each of the output variables.

4. *Defuzzification of the Results*: The fuzzy sets from the previous step are transformed into output crisp values. There are multiple techniques to carry out this process, among which the Middle of Maximum (MoM) and the Center of Area (CoA) stand out.

5. A Model of Virtual Characters Using Fuzzy Logic

As described previously, the aim of this work is to build a model of virtual characters that can simulate emotions caused by the events that occur in a virtual environment, using fuzzy logic (see Section 4) and computational modeling equations. In addition, this model provides components that can filter emotions in order to obtain a more complex emotional state through the use of the character's mood and personality. In this section, we describe how this model has been designed.

The proposed model takes the computational model used in FLAME (El-Nasr et al., 2000, see Section 3.2) as a starting point in order to obtain the emotions according to the events that take place in the virtual environment. These events can occur either when a character interacts with an object with another character. In the first case, there must be some relation between the character and the object so that an emotion can arise (e.g., a character gets angry when another character takes its beverage, as it may understand that the beverage is being stolen). If there is no relation between the character and the object, then the event will not have any emotional impact.

In the following sub-sections, we describe the main features and mechanisms used to define a computational model of emotion and the relationships that exist between them to determine an emotional state in response to an event. Subsequently, we explain how we have integrated mood and personality into this model, in order to create a richer emotional model of virtual characters.

5.1 Emotions

According to the FLAME model, the agents that inhabit a virtual environment receive information from the events that occur in the environment through the agent's perceptions, which are then sent to their emotional component. This element processes perceptions and uses the outputs of an inductive learning component (that calculates the expectation of an event) in order to simulate an emotional state. In the proposed model, we have substituted the learning component with one based on the frequency of events, as the number of events that are likely to happen is restricted and a learning-based element is not likely to provide good results. In our case, the expectation of an event will be calculated according to its frequency—the more frequently it occurs, the higher the expectation of that event will be. This complies with the role of the expectation of an event, which reduces the intensity of the emotions it causes when an event repeats frequently.

The first process of the emotional component is the event evaluation (Fig. 2), where the system captures and processes the environment variables when the event occurs. In the proposed model, this evaluation is made in two steps—in the first one, the goals affected by the event are identified, along with the impact that the event has on them; in the second one, the rules to obtain the desirability of the event are calculated according to the impact calculated in the previous step and the importance of the goals. Therefore, the event evaluation process describes three variables that determine the emotional state:

- **Importance:** This variable expresses the importance of the character's goals affected by the event.
- **Impact:** This variable represents the impact of the events on the character's goals.
- **Desirability:** This variable shows the level of desirability of the event according to the importance and impact on the character's goals.

All the characters endowed with emotions have goals with a certain associated importance for the character. During the evaluation process, once the affected goals have been identified, their corresponding importance is fuzzed, using fuzzy sets, according to the precision desired for the classification of the corresponding crisp values. However, it is also necessary to consider that the inference process will turn more complex if the set granularity increases. In this case, we have empirically determined that in order to classify the event importance, it is enough to use just three sets: Not Important, Slightly Important and Extremely Important (Fig. 5).

After calculating the importance of the event, the *impact* degree that the event causes on the goals is set. Given the significance of the character's goals on the model, it is necessary to use a higher precision to fuzzily the impact than in the case of the importance of the event. In this case, it has been empirically determined that using five fuzzy sets allows us to model the impact with enough precision: *Highly Positive, Slightly Positive, No Impact, Slightly Negative* and *Highly Negative* (Fig. 6).

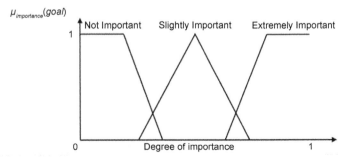

Fig. 5: Membership functions for the variable importance.

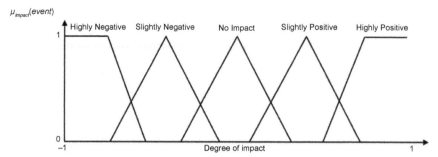

Fig. 6: Membership functions for the variable impact.

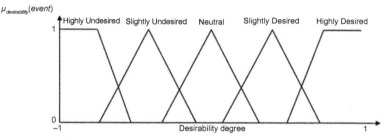

Fig. 7: Membership functions for the variable desirability.

Finally, the level of *desirability* of the event from the character's point of view is established according to the values of the *impact* and the *importance* we have just calculated. The possible values for the desirability are *Highly Undesired, Slightly Undesired, Neutral, Highly Desired, Slightly Desired* (Fig. 7). The value of the desirability is established using a set of fuzzy rules (Section 4.2) like the following ones:

IF Impact (G_1,E) is A_1
AND Impact (G_2,E) is A_2
...
AND Impact (G_k, E) is
A_k AND Importance (G_1)
is B_1 AND Importance
(G_2) is B_2
...
AND Importance (G_k) is B_k
THEN Desirability (E) is C

where Gi are goals and E is the event that has triggered the rules. Applying the membership functions of the impact and the importance, we obtain fuzzy sets that can be trivially matched in the rule antecedents with the values of the different Ai and Bi. If all the antecedents hold, the set C is associated with the desirability of the event E (i.e., if the event E has an impact Ai on the goals Gi and each goal Gi has an importance Bi, then the desirability of the event E is C).

If we have a character that needs to send an urgent text message and another character steals its smartphone, this event may have a negative impact on the goals of the first character, so it can be easily seen that the event is undesirable. A fuzzy rule that defines this situation is

> IF *Impact* (send text message, stolen phone) is *Highly Negative* AND *Importance* (send text message) is *Extremely Important* THEN *Desirability* (stolen phone) is *Highly Undesired*

In order to create and process the rules complying with the properties of the classical sets and at the same time allow their use in an easy way, we have chosen the Mandani inference model (Section 4.2). This method allows us to create IF-THEN rules and nest additional antecedents in order to create more complex rules. The simplicity of the well-defined steps of this model makes it easy to obtain a final crisp value, using different fuzzification and defuzzification methods.

In the current model, we have decided to use the Middle of Maximums (MoM) method to defuzz the value of the desirability of an event. Provided that we are only using five basic emotions for the characters (joy, sadness, fear, anger, and surprise), we have considered that it is precise enough for our purposes, even though other methods, such as the Centroid, may be more acute (and computationally more expensive). The result of the defuzzification process is a crisp value that represents the desirability of an event, which is subsequently used to calculate (using the formulas proposed in the subsequent sections) the intensity of the emotions of a character that are triggered by that event.

5.1.1 Emotion Classification

Once the expectation of an event has been identified and all the rules have been processed, and therefore, the fuzzy set corresponding to the desirability of the events has been processed, we can calculate the emotional state of the agent, using affective computing equations, which depend on both the expectation and the desirability of the events. The source of these equations is in the relation between the emotions and the desirability of an event, based on the definitions provided by Ortony et al. (1990), as shown in Table 2.

The proposed model considers both the desirability and expectation of an event as variables needed to calculate the intensity of emotions. However, variables, such as the implication or the approval of other characters, are beyond the scope of this work. Therefore, in this work we use some of the emotions proposed in FLAME, which do not depend on social norms, but just on the character's own actions and decisions: joy, sadness, fear, anger and surprise (Table 3).

The emotions of joy, sadness, and fear have been modeled according to the equations proposed by Price et al. (1985) and adapted so that they can be used in terms of the desirability and expectation of an event.

However, anger as a complex emotion depending on a combination of sadness and reproach, is defined as an action disapproved by other characters. A reproachable action could be stealing a cellphone, which would make all the characters angry, and not only the one owning the stolen phone. As we are not considering reproach, our current proposal is to consider anger as a combination of sadness and fear.

Surprise is an emotion that is not included in FLAME, but we have decided to consider it in our model in order to complete the set of basic emotions proposed by the discrete theories of emotion (Section 2.1).

Table 2: Rules for modelling emotions in FLAME.

Emotion	Rule
Joy	Occurrence of a desirable event
Sadness	Occurrence of an undesirable event
Disappointment	Occurrence of a disconfirmed desirable event
Relief	Occurrence of a disconfirmed undesirable event
Hope	Occurrence of an unconfirmed desirable event
Fear	Occurrence of an unconfirmed undesirable event
Pride	Action carried out by the character approved by other characters
Shame	Action carried out by the character disapproved by other characters
Reproach	Action carried out by other character disapproved by this character
Admiration	Action carried out by other character approved by this character
Anger	Complex emotion: sadness + reproach
Gratitude	Complex emotion: joy + admiration
Gratification	Complex emotion: joy + pride
Remorse	Complex emotion: sadness + shame

Table 3: Theoretical foundation for modeling emotions.

Emotion	Rule
Joy	Occurrence of a desirable event
Sadness	Occurrence of an undesirable event
Fear	Occurrence of an unconfirmed undesirable event
Anger	Complex emotion: fear + sadness
Surprise	Occurrence of an unconfirmed, desirable or undesirable event

5.1.2 Equations for the Intensity of Emotions

After establishing the theoretical foundations to model emotions, it is possible to create the equations that calculate the intensity of emotions in terms of the desirability and the expectation of an event, obtained when processing the rules that model the emotions. The original equations were designed by Price et al. (1985) (Table 4) and each of them uses a pre-calculated coefficient that provided good experimental results.

In the proposed model, we had to test these equations and adjust the values of the coefficients in order to obtain a behavior that was closer to what we expected. In the case of the emotions not considered by FLAME, such as surprise, or with undefined equations, such as anger, the proposed equation has the same structure of the existing formulas, as shown in Table 5.

5.1.3 Emotion Inhibition

Once the emotional state has been determined using the previous equations, it is necessary to establish how some emotions affect others. For example, sadness is usually affected by

Table 4: Intensity of emotions in the FLAME model.

Emotion	Equation
Joy	(1.7 × expectation 0.5) + (−0.7 × desirability)
Sadness	(2 × expectation 2) − desirability
Disappointment	Hope × desirability
Relief	Fear × desirability
Hope	(1.7 × expectation 0.5) + (−0.7 × desirability)
Fear	(2 × expectation 2) − desirability

Table 5: Intensity of emotions in the proposed model.

Emotion	Equation
Joy	(1.2 × expectation 0.5) + (−0.6 × desirability)
Sadness	(2 × expectation 2) − desirability
Fear	(2 × ((1 − expectation)/4) 2) − desirability/2
Surprise	|expectation| × (1 + desirability)
Anger	(Fear × Sadness) + expectation/2

emotions, such as anger or fear (Bolles and Fanselow, 1980). In general, emotions with a higher intensity tend to influence the rest; so a filtering process is needed in order to calculate the threshold over which an emotion starts influencing others.

The relationship between two or more emotions that can suppress other emotions is known as inhibition. Some models of emotions use techniques in which weak emotions opposed to the more intense ones are inhibited (Velásquez, 1997). For example, sadness would tend to inhibit happiness if it had a higher intensity. In the current model, opposite emotions are inhibited according to their intensities, but negative emotions (fear, anger, and sadness) are preferred over positive ones (joy) as they tend to be more persistent and dominant.

The main inconvenience when using these techniques is deciding when an emotion should not influence another one. For example, when a series of events where sadness and joy change continuously, as soon as the intensity of one of them exceeds the intensity of the other, then the second emotion would be inhibited, and this alternation could last forever. This conflict can be solved by defining an error range, so that when opposing emotions have similar intensities, then inhibition is not possible. In case the difference between the intensities exceeds the error range, then the more intense one can inhibit the other. However, the real problem is that we are not taking into account the big picture of the whole emotional state, as this inhibition may not be reasonable.

In the following Sections (5.2 and 5.3), we propose a solution to this problem by taking into account the mood of the character and, subsequently, we show the influence of other factors, such as the character's personality in order to model a more realistic emotional response.

5.1.4 Emotion Decay

The last step to simulate emotions is to use a method that allows us to fade the emotions of the agents in the absence of new events. When the stimulus that triggers an event disappears, the emotion does not vanish instantly; instead, it decays with time. This process of emotional decay can be modeled in different ways. In the current model, we propose the use of two constants to define how both positive and negative emotions decay. For positive emotions, we define a constant φ, and for the negative ones a constant ϕ, such that $\varphi < \phi$. This way, similarly to what usually happens with humans, negative emotions tend to be more persistent than positive ones, which decay faster. Emotion decay is progressive until their value goes down to 0.

5.2 Mood

At this stage, the mood can help us filter the interactions between the different emotions in order to solve the inhibition problem. Mood can be represented as a summary of past emotions by means of a factor that can be either positive or negative. This value depends on the intensity of the character's emotions, either negative (sadness, rage or fear) or positive (joy).

When a new event is triggered, the intensities of the different emotions are likely to vary, as they are not constant over time, but just using the last emotional state of a character in order to decide whether it is in a good or bad mood may be misleading. As an example, let's imagine two characters, David and Laura, who are walking along the river shore. Suddenly, a thief steals David's wallet. David starts chasing the thief, but a bicycle runs over him and makes him fall into the river. Laura helps David get out of the water and returns him his wallet, which the thief lost while he was running away.

While all these events were happening, David's mood was worsening with every new mishap. If the mood only takes into account the last emotional state generated by the emotional component, then when the story finishes, David should be happy because he managed to retrieve his wallet, forgetting about his stolen wallet, the bicycle running over him and falling into the river, which would not be very realistic. Instead, the mood component should take all these events into account, over a time-window of a certain length, in order to generate a more believable mood value.

5.2.1 Time-Window Definition

The solution we have adopted to solve this issue is to calculate the mood value using the emotions generated over the last n time units. We have heuristically determined that using the last 5 time slots offers the most coherent results, since using a longer time frame makes the mood consider events which are not recent enough. Similarly, using a shorter time frame makes the mood too similar to the case in which only the last event is considered. The equations used to calculate the mood value are:

$$mood = \begin{cases} positive & \text{if } \sum_{i=-n}^{-1} I_i^+ > \sum_{i=-n}^{-1} I_i^- \\ neutral & \text{if } \left| \sum_{i=-n}^{-1} I_i^+ \right| - \left| \sum_{i=-n}^{-1} I_i^- \right| > 10\% \\ negative & \text{if } \sum_{i=-n}^{-1} I_i^+ < \sum_{i=-n}^{-1} I_i^- \end{cases}$$

where n is the number of time slots (i.e., the size of the time-window), I+ is the intensity of the positive emotions in instant i and I− is the intensity of the negative emotions in instant i.

To illustrate how the calculation of the mood works, we can use the following example:

The current mood has a negative value as a result of three negative emotions (sadness, anger and fear) versus a positive one (joy), all of them with a medium intensity. An event is triggered and the new emotional state shows that joy has an intensity of 0.45 and anger has an intensity ≤ 0.20. In this case, anger inhibits joy even though the intensity of joy is greater than the one of anger, since the mood of the character at that moment was negative.

5.2.2 Emotion Filtering Using Mood

In order to prevent the mood to always be influenced by the intensity of the emotions, we must define a tolerance value between the positive and negative emotions. For example, in FLAME, this value is 5 per cent, but it can be any value that adjusts the behavior of the system providing coherent results. In this case, if an emotion has a value v, any emotion with a value $\pm$ 5 per cent will depend on the mood in order to obtain its final intensity.

In the proposed model, the result obtained at the end of this process is a number that indicates whether the mood is positive, neutral or negative. This value is then used to filter the emotions obtained after processing the rules that model them, running Price's equations and getting the emotional state. In the case of positive emotions, when a new event is triggered, the filtered emotion is calculated according to the following formula:

$$I_m^+ = \begin{cases} I^+ - (mood \,/\, windows \times desirability) & \text{if } desirability < 0 \\ I^+ + (mood \,/\, windows \times desirability) & \text{otherwise} \end{cases}$$

where I+ is the intensity of a positive emotion after filtering it with the mood, and I+ is the intensity of the resulting emotion after using the emotion equations. The value of the mood is divided by the maximum value that can be obtained by subtracting the summation of the positive and negative emotions (number of time-windows). If, for example, all the negative emotions of the time-windows had a value 0.0 and the positive ones had a value of 1.0, then the resulting intensity would only depend on the desirability and the current intensity. In case the moods were neutral, the emotions would be processed as if there was no mood.

In order to calculate the negative emotions, we proceed in a similar, but subtracting the value of the intensity if the event is desirable:

$$I_m^- = \begin{cases} I^- - (mood \,/\, windows \times desirability) & \text{if } desirability > 0 \\ I^- + (mood \,/\, windows \times desirability) & \text{otherwise} \end{cases}$$

The only emotion that is not comprised within either positive or negative emotions is surprise. Offering an unexpected gift or bringing bad news are opposite actions where surprise might reach high levels, so none of the previous equations is valid for surprise. On the contrary, we need more variables apart from desirability to determine whether the intensity is going to increase or decrease. The filtering function for surprise is defined as follows:

$$I_m = \begin{cases} I_s - (mood \,/\, windows \times desirability) & \text{if } desirability < 0, mood < 0 \\ I_s + (mood \,/\, windows \times desirability) & \text{if } desirability < 0, mood > 0 \\ I_s + (mood \,/\, windows \times desirability) & \text{if } desirability > 0, mood < 0 \\ I_s - (mood \,/\, windows \times desirability) & \text{if } desirability > 0, mood > 0 \end{cases}$$

Considering the current mood, we can define 4 different cases of opposed situations to calculate the intensity of surprise. For example, if Mary gives away a smartphone to Carl, who is not in a good mood because he has dented his car door when parking it, his surprise will be remarkable because Carl would not be expecting something positive to happen. On the contrary, if Mary gives Carl a parking manual, he will be in a worse mood and surprise will not increase nearly as much as his negative emotions.

5.2.3 Integrating Short-term Memory

In the proposed model, the mood captures the last time-windows of the emotional state in order to determine whether it is positive or negative and, according to this value, it recalculates the equations of the emotional modeling. In addition, the proposed mood filtering equations (Section 5.1.3) manage to simulate emotions according to what has happened in the last n time-windows. However, the mood currently only uses the obtained value as a general influence of the emotional state in previous moments. This can make the system behave as if it were only reacting to isolated events, obtaining incoherent results.

In order to clarify how the mood works, we are going to illustrate it through an example. Let's suppose a simulation where the mood of all the characters is neutral, and we have a character named Jester, who is in a canteen. Jester steals Robert's cellphone, where Robert is another character, who is standing close to the bar. Immediately after that, Jester also decides to steal Robert's beverage.

When Jester steals the cellphone, Robert's mood moves from neutral to negative, provided that in the emotional state, calculated in each time-window, the intensity of negative emotions is higher than the one of positive emotions. During this process, the mood has not been taken into account to obtain the value of emotions, since the character initially had a neutral mood and the subsequent filtering process has no effect. On the contrary, when Jester steals Robert's beverage, the emotions are calculated by using a negative mood; so the negative emotions increase more than when the mood is neutral.

However, Robert considers his cellphone more important than his beverage. If, after the first robbery, the intensity of anger is 0.70 and sadness and fear have both a value of 0.50, the mood will be quite negative when the second robbery is processed. If the beverage is less important than the cellphone, then the intensity of anger starts at 0.20 and filtering of the mood results in a value which might not exceed 0.30.

As a result, after the first robbery, the character's anger has a value of 0.70, and after the second one, the value is only 0.30 (i.e., after the second robbery the character is less angry than after the first; he may be incoherent even though the importance of the beverage is less than that of the cellphone). This effect makes it look as if the moods were not taking into account all the previous actions and reacted independent of each event. If the events had happened in the inverse order, this error would not have been perceived as the intensity of the negative emotions after the second robbery is higher than after the first.

In summary, the mood alone is not enough to obtain coherent results. So we need to consider the history of previous emotions and events, storing them in what is called a *short-term emotional memory* (Ortony et al., 1990) integrated in the mood. The purpose of this new element is to aggregate the intensities of the previous emotions to the current mood, so that the effect of previous actions on the current mood can be perceived. In emotions such as surprise, fear and joy, we have used the average of the intensity before the event is triggered and after the mood has been filtered. Sadness and anger are persistent negative emotions. So their calculation is a little more complex. In the case of sadness, the final intensity is the maximum of the two intensities. The equation for anger also depends on the expectation, and is as follows:

$$max \ (new \ anger, \ old \ anger) + min \ (new \ anger, \ old \ anger) \times expectation$$

5.3 Personality

Use of the equations defined by Price et al. (1985) is the basis for emotional modeling of the characters. The use of mood allows us to model the emotions considering the previous events that the character had experienced, preventing their reactions to specific events to be always the same, irrespective of what had happened previously. However, using the current model, all the characters will behave in the same way, as there is nothing to allow them to behave differently. The element that can allow us to solve this issue is personality, as characters with different personalities will react differently to the same event. In this section, we are going to describe how we have integrated the Big Five personality model (Barrick and Mount, 1991) to add a little more complexity and realism to the emotional simulation of characters.

5.3.1 Mapping Personality Traits

In the existing literature, the *Big Five* model is integrated with moods through a value mapping that fits the PAD model (Section 2.2). In our case, we are not using the PAD model but fuzzy logic, as it offers a wider range of possibilities to represent emotions instead of just a fix set of values. In addition, we have used FLAME as a reference to model and integrate both emotions and mood, but it does not consider the personality traits of an agent. In this case, we have mapped the emotions with the personality traits described in the *Big Five* model, using the descriptions provided by ALMA (Gebhard, 2005) (Fig. 8).

The result of this mapping can be seen in Table 6, where each personality trait has both a high and a low value matching the level of intensity of the personality trait and how the character's emotions are affected by that trait, depending on their intensity. For example, an agreeable character tends to be happy and calm, while a low level of agreeableness makes a character angry and sad. The emotions that appear in the high-level and low-level sets of a personality trait may not be completely opposed to one another, and it is possible not to have the same emotions in both sets, as in the case of agreeableness.

This mapping is used to obtain the other factors, such as the moods, which increase or decrease the intensity of emotions. These values are fixed and do not change over time, as personality does not change with time, in contrast to what would happen with moods.

The input to the process of filtering the emotions with the personality traits are the emotions that have already been already filtered, using the mood. After that, the

Trait	Refers to	If low score
Openness	Imaginative, prefer variety, independent	Down-to-earth, conventional, low aesthetical appreciation
Conscientiousness	Well-organized, careful, reliable, self-discipline	Disorganized, careless, weak-willed
Extraversion	Sociable, affectionate, optimistic	Reserved, sober
Agreeableness	Trusting, helpful	Suspicious, cynical
Neuroticism	Anxiety, experience negative emotions, vulnerable	Secure, calm, self-satisfied

Fig. 8: Personality traits in ALMA (Gebhard, 2005).

Table 6: Mapping of personality traits and emotions.

Traits	High Level	Low Level
Openness	++joy, –surprise, –fear, –anger, –sadness	+surprise, –joy, +fear
Conscientiousness	–surprise	+surprise
Extraversion	+joy, –fear, +anger	+fear, –anger
Agreeableness	+joy, –anger	–anger, –sadness
Neuroticism	+anger, +fear, +sadness, +surprise	–anger, –fear, –sadness, –surprise

personality influences the emotions according to the values that appear in Table 6. Each positive or negative sign of emotion in the table represents a change in the value of that emotion by a certain value δ. For example, a person with a high level of openness will increase $2 \times \delta$ the value of joy with respect to the value it had after filtering it with the mood. We have determined empirically that a value of δ in the range of 10–15 per cent offers coherent results according to what different users expected in testing the model.

6. Implementation of the Model

This section describes the methods adopted to build the model proposed in the previous section, using the Unity 3D video game engine. The project is developed in the programming languages, C# and XML; this last one is used for loading configuration files.

The architecture of the emotional agent (Fig. 9) is composed of three main components: emotions, mood, and personality. The agent communicates with the

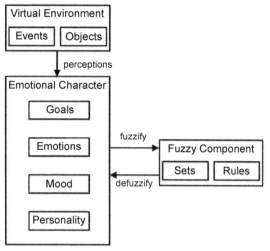

Fig. 9: General view of the system architecture.

fuzzy component, which contains the rules and sets defined by the user to calculate the desirability of the event in order to obtain the emotional state. Next, the implementation of how each component has been carried out will be explained.

6.1 Virtual Environment

The virtual environment contains information that is generated while the agent explores the environment or represents facts known by all agents. It's used as input of the simulating emotions process in virtual agents and can be classified in two groups of information about the environment: objects and events.

The objects are represented in Unity through *Game Objects*, which represent containers of all the components that implement the real functionality. All the objects store information that allows defining a relationship between a certain element of the environment and a character, through the importance it has for the character.

Fig. 10: General view of the virtual environment.

Application events can be produced in two ways, using behavioral scripts in objects and characters of the virtual environment, when an action is performed on an object of the environment or during the interaction with another character. In the case of actions, there must be a relationship between the object and the character to simulate an emotion about it. If there is no character related to the object, the event will have no effect on the application.

In this system, the expectation begins at a minimum value and increases as the same event is repeated until it reaches the maximum value. This means that the event has occurred so many times that the impact on the character will be minimal. An example that we can find in the application is the theft of an object. The player steals a soda from a character and then his mobile. The second robbery will have a greater expectation because it has already experienced it before, which causes emotions like surprise to be seriously depleted.

6.2 *Message Controller*

Asynchronous messages allow the instructions of a function to not be blocked by waiting for the results of previous operations. This is very beneficial in case that several events happen consecutively and it is necessary to execute operations, regardless of the order and without affecting the final result. Therefore, all environment elements communicate the event they produce, using the asynchronous message-passing mechanism provided by Unity.

The message is sent to a function of the program to be executed and the receiver searches among all its behavior scripts if there is a function that corresponds to the requested one. In case of not finding it, the event won't be processed. On the other side, the indicated function will capture the message while the flow of operations invoked by the event continues its course.

The *camera* is another object of the game that captures and shows the virtual world created in Unity during the simulation. The execution of certain events may change the user interface, for example, in giving feedback to the user of what is happening. These changes are updated with the asynchronous messages mechanism, so that the camera communicates with the message controller, which is responsible for managing the reception of messages.

When an action related to an object is executed, the message controller obtains the identifier of the owner or the person involved in the event to know which character to send an asynchronous message, with all the information of the event and the object. At the end of the process, the object disappears from the scene. In this way, possible conflicts are avoided in case the player tries to repeat the action several times because the event expectation will reach the maximum value, leaving the owner of the object completely insensitive to the event.

The resending of messages ends in the emotional component, which is responsible for generating the final emotional state. Each virtual agent has a script that allows simulating emotions and containing the current state of the intensity of emotions, the mood as a multiple emotional states, the goals, and the personality traits.

6.3 *Emotional Component*

Each agent keeps its own objectives within the behavior scripts, following a data structure similar to the events. The objectives of each agent do not belong to the environment

Fig. 11: Depiction of the character's emotions.

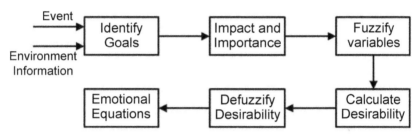

Fig. 12: Flow of the simulation of emotions.

information, which is shared between all the agents but depends on each one to manage their own objectives.

The emotional component is the basis for the simulation of emotions to obtain a basic emotional state. Therefore, it is a required component for the good performance of the model but it doesn't need any initial configuration by the user because all the operations that calculate the intensity of emotions are predefined and described in Section 5.1.

Figure 12 shows the process of the emotional component, which receives the event and the information of the object involved as the input. First, the character goals affected by the event are identified and the event expectation is increased. Next, the rules manager of the fuzzy component generates the variables of impact and importance of the goals based on the acceptance degree of the event, the goals weight, and the importance of the object.

Then, they are *fuzzed* in order to process the rules and obtain the fuzzy set of desirability, which is immediately *defuzzed* to use a crisp value and calculate the emotional equations of Section 5.1.2. In this way, we get the most basic emotional state that this model can simulate.

Finally, the decay of emotions continues as time goes by. In this model, it has been decided that the intensities decay every second during the simulation. Unity provides the static variable *DeltaTime* that takes the time in seconds to complete a frame. So every second the emotions intensities will decrease, using decay constants for positive and negative emotions (Section 5.1.4).

Fig. 13: Depiction of the character's emotions and mood.

6.4 Mood Component

The model of characters can add the mood component to make the emotion simulation process more complex. The incorporation of the mood is optional to allow the user to decide if he only uses the emotional equations or combines both components to get a more realistic emotional state.

To carry out this task, the model allows the use of configuration files in a markup language so standardized and easy to use for anyone, such as XML. In this way, the user can easily decide to incorporate this feature through the configuration described in the corresponding files.

There are many ways to parse an XML file but in Unity the most practical one is to use the *XML Serializer* library, available with languages, such as C#, .Net and Javascript. This library allows reading operations of XML annotations and communicates to the container how the file should be parsed.

6.4.1 Monitoring of Emotional Intensities

Once the configuration is established and as soon as the system execution begins, it will be loaded together with the environment data (goals of each character, objects, and possible events). From that moment, the mood will begin to store the most recent emotional states to calculate the difference between positive and negative emotions. The continuous update of the intensities ensures that the simulation of emotions is always conditioned by the mood. To avoid this problem, we use a margin of error between the positive and negative emotions in which the mood doesn't perform. A tolerance value of 10 per cent is taken because of its good results during the simulation.

Section 5.2.1 explained the importance of choosing the right number of time-windows to capture the intensities of emotions. The value that gave the best results in Unity was 15; so it was used as the maximum number of windows.

Finally, we implemented the short-term memory as an extension of the mood. After the calculation of the new emotions and applying the filtering, short-term memory is responsible for requesting the last emotional state and combines it with the old intensities according to the equations of Section 5.2.3. And this is the last step to achieve the final emotional state of mood processing.

Fig. 14: Depiction of the character's emotions, mood and personality.

6.5 *Personality Component*

The personality is treated as a component that stores the personality traits of an agent. These features are predefined by the user through XML files before the simulation starts and lasts until the end of it. The input data received by the personality component are the intensities generated by the emotional equations or, if the mood is enabled, those obtained after this phase. The output is the emotional state that combines personality traits with emotions.

The incorporation of the personality to the behavior of a character is optional, as it was with the mood. At the end, the modeling of emotions would be even more complex because the results will be much more coherent than they were until now.

The personality of an agent is immutable, so from the initial charge, the character has the same value for each trait of the Big Five until the end of execution. Each of them serves as a factor to increase or decrease the intensities of mood processing, multiplying the factor by the intensities. Its increase or decrease addresses the descriptions in Table 6 and varies from 15–30 per cent as is explained in personality traits mapping (Section 5.3.1). And this is the last step of the simulation of emotions in the most complete way, combining emotions with mood and personality.

7. Conclusion

Our research on computational models and psychological theories of emotion culminated in the adoption of the FLAME computational model as a basis for the process of generating emotions. It was chosen mainly for its model of emotions based on fuzzy logic techniques, but its processing remains stuck when combined with the concept of mood, since the underlying model continues to respond independent to the events. In this chapter, a model of emotion simulation for virtual characters has been described, using fuzzy logic for emotion simulation and incorporating new elements to enrich it with mood and personality traits.

The mood solves the problems that some models encounter when a character forgets what has happened immediately before. It also adds new features to improve the emotional state of the output. Regarding personality traits, we have chosen to rely on the model that has become a psychological standard: the *Big Five*. The model of the *Big Five* addresses the personality from the five-dimensional point of view and the descriptions of the emotions that were used to formulate an approach to the human personality.

Once the theoretical design of the model was established, we integrated it into humanoid characters within a virtual environment created in Unity 3D. It has been implemented mainly by separating the emotional, personality and mood components, so that different combinations can be made, separating the personality or mood from the emotion-generation process.

After performing some tests to check how the system behaved in Unity 3D, both expert and inexperienced users were asked to perform evaluations to check the realism of the model and assess the coherence of the generated emotions. Given the results of the evaluation, it is worth mentioning that if the mood and personality had not been implemented, the results obtained would have been much worse than expected.

In the current implementation, it is still pending to show the resulting emotions, both in the facial expression and the body posture of the characters, to better reflect in the 3D environment what the emotional state of the characters is.

References

Allport, G.W. (1937). Personality: A Psychological Interpretation, Holt.

Barrick, M.R. and Mount, M.K. (1991). The Big Five personality dimensions and job performance: A meta-analysis. Personnel Psychology, 44(1): 1–26.

Bee, N., Pollock, C., Andre, E. and Walker, M. (2010). Bossy or wimpy: Expressing social dominance by combining gaze and linguistic behaviors. *In*: International Conference on Intelligent Virtual Agents, pp. 265–271, Springer.

Bolles, R.C. and Fanselow, M.S. (1980). A perceptual defensive recuperative model of fear and pain. Behavioral and Brain Sciences, 3(2): 291–301.

Cattell, R.B. (1957). Personality and Motivation Structure and Measurement. World Book Co.

Damasio, A.R. and Sutherland, S. (2008). Descartes' error: Emotion, reason and the human brain. Nature, 372(6503): 287–87.

Dias, J., Mascarenhas, S. and Paiva, A. (2014). FAtiMA Modular: Towards an Agent Architecture with a Generic Appraisal Framework, pp. 44–56, Springer International Publishing.

Ekman, P. (1992). An argument for basic emotions. Cognition and Emotion, 6(3-4): 169–200.

El-Nasr, M.S., Yen, J. and Loerger, T.R. (2000). Flame-fuzzy logic adaptive model of emotions. Autonomous Agents and Multi-agent Systems, 3(3): 219–257.

Eysenck, H.J. (1991). Dimensions of personality: 16, 5 or 3? Criteria for a taxonomic paradigm. Personality and Individual Differences, 12(8): 773–790.

Forgas, J.P. (1995). Mood and judgment: The affect infusion model (aim). Psychological Bulletin, 117(1): 39.

Frijda, N.H. (1986). The Emotions. Cambridge University Press.

Frijda, N.H. et al. (2000). The psychologists' point of view. Handbook of Emotions, 2: 59–74.

Gebhard, P. (2005). Alma: A layered model of affect. *In*: Proceedings of the Fourth International Joint Conference on Autonomous Agents and Multiagent Systems, pp. 29–36, ACM.

Hudlicka, E. (2011). Guidelines for designing computational models of emotions. International Journal of Synthetic Emotions (IJSE), 2: 26–79.

Klir, G. and Yuan, B. (1995). Fuzzy Sets and Fuzzy Logic. vol. 4, Prentice Hall, New Jersey.

Lazarus, R.S. (1991). Emotion and Adaptation. Oxford University Press (on demand).

Mairesse, F. and Walker, M.A. (2010). Towards personality-based user adaptation: Psychologically informed stylistic language generation. User Modeling and User-Adapted Interaction, 20(3): 227–278.

Marsella, S., Gratch, J. and Petta, P. (2010). Computational Models of Emotion. Chap. 1.2, pp. 21–46, Oxford University Press.

Marsella, S. and Gratch, J. (2014). Computationally modeling human emotion. Communications of the ACM, 57: 56–67.

Marsella, S.C. and Gratch, J. (2009). Ema: A process model of appraisal dynamics. Cognitive Systems Research, 10(1): 70–90.

McRorie, M., Sneddon, I., McKeown, G., Bevacqua, E., de Sevin, E. and Pelachaud, C. (2012). Evaluation of four designed virtual agent personalities. IEEE Transactions on Affective Computing, 3(3): 311–322.

Mehrabian, A. (1995). Framework for a comprehensive description and measurement of emotional states. Genetic, Social, and General Psychology Monographs.

Ortony, A., Clore, G.L. and Collins, A. (1990). The Cognitive Structure of Emotions. Cambridge University Press.

Pourjavad, E. and Mayorga, R.V. (2017). A comparative study and measuring performance of manufacturing systems with Mamdani fuzzy inference system. Journal of Intelligent Manufacturing, pp. 1–13.

Price, D.D., Barrell, J.E. and Barrell, J.J. (1985). A quantitative experiential analysis of human emotions. Motivation and Emotion, 9(1): 19–38.

Reisenzein, R., Hudlicka, E., Dastani, M., Gratch, J., Hindriks, K., Lorini, E. and Meyer, J.-J.C. (2013). Computational modeling of emotion: Toward improving the inter- and intra-disciplinary exchange. IEEE Transactions on Affective Computing, 4: 246–266.

Roseman, I.J., Spindel, M.S. and Jose, P.E. (1990). Appraisals of emotion-eliciting events: Testing a theory of discrete emotions. Journal of Personality and Social Psychology, 59(5): 899.

Strongman, K.T. (2003). The Psychology of Emotion: From Everyday Life to Theory, 5th ed., John Wiley & Sons.

Sun, R. (2008). Introduction to Computational Cognitive Modeling. Cambridge Handbook of Computational Psychology, pp. 3–19.

Velasquez, J.D. (1997). Modeling emotions and other motivations in synthetic agents. pp. 10–15. *In*: Proceedings of the Fourteenth National Conference on Artificial Intelligence and Ninth Conference on Innovative Applications of Artificial Intelligence, AAAI'97/IAAI'97, AAAI Press.

Chapter **6**

Pedagogical Agents as Virtual Tutors
Applications and Future Trends in Intelligent Tutoring Systems and Virtual Learning Environments

Hector Rafael Orozco Aguirre

1. Introduction

In the situation of a classical lecture or seminar, adaptation and personalization can take place as well. It means that the teacher adapts his or her way of teaching or adapts the content, based on feedback received by the students. Pivec et al. (2004) specified two possibilities for this feedback to be provided: explicit feedback and implicit feedback. Explicit feedback means that the students convey to the teacher what they have understood from the learning content, what other content they would like to know, and what other kind of teaching they would like to have. Implicit feedback is information that the teacher receives from the students' verbal signals during or after the lecture or seminar, such as noise level or nodding with the head.

As was introduced by Imbert et al. (2007), the use of virtual reality (VR) and virtual environment (VE) technology combined with intelligent tutoring systems (ITSs) proved a valuable and promising approach to computer-based teaching and training. Together with the traditional properties of ITSs instruction, personalization of the contents and presentation and adaptation of the tutoring strategy to the student needs, VR allows the student to gain practice and skills by interacting with virtual scenarios similar to the real ones, but without assuming any of their potential risks. Also, as not only the trainee but also the tutor may have a virtual representation (virtual mannequin or *avatar*) in the 3D environment, it can be used to perform demonstrations about how to carry out specific tasks, just in the same way in which a human tutor may proceed with his students. Taub et al. (2016) argued that VR has attracted more and more interest as one of the

Autonomous University of Mexico State, Mexico.
Email: hrorozcoa@uaemex.mx

most potential propositions to change and improve education. Previous research works have shown that this new technology seems to have a positive influence on learning in educational applications. Besides, the presence of interactive virtual agents, also called 'embodied conversational agents', by taking the role of tutors, seems to have a positive effect on student engagement and the effectiveness of teaching.

According to (Baker, 2016), one of the initial visions for intelligent tutoring systems was a vision of systems that was as perceptive as a human teacher and as thoughtful as an expert tutor, using some of the same pedagogical and tutorial strategies as used by expert human tutors. These systems would explicitly incorporate knowledge about the domain and pedagogy, as part of engaging the students in a complex and effective mixed-initiative learning dialogues. Student models would infer what a student knew, and the student's motivation, and would use this knowledge in making decisions that improved student outcomes along multiple dimensions. An intelligent tutoring system would not just be capable of supporting learning, but behave as if it genuinely cared about the students' success. These systems would not only be effective at promoting students' learning, but the systems themselves would also learn how to teach. After decades of hard work by many world-class scientists, there are wonderful demonstrations of the potentials of this type of technology (Kulik and Fletcher, 2016). There exist systems that can provide support at every step in a student's thinking process; systems that can talk with students in natural language; systems that model complex teacher and tutor pedagogical strategies; systems that recognize and respond to differences in student emotion, and simulated students that enable human students to learn by teaching (Almasri et al., 2019).

Interactive agents used for pedagogical purposes induce a higher motivation to learn in students. These agents, known as 'pedagogical agents', enhance learning by attracting attention through a dynamic interactive and conversational mechanism with students, through a bidirectional dialogue. Pedagogical agents are seen as anthropomorphic virtual characters mainly used for educational or training purposes (Azevedo et al., 2016). The design of pedagogical agents has changed over time, depending on the desired objectives through their use. This field of research has emerged over the past three decades to provide support and achieve an adequate teaching-learning process, as per the student and teacher needs. Apart from being able to communicate verbally, it is desirable that created pedagogical agents are full-bodied, and can show emotions or facial expressions and even body motions to be able to communicate in a non-verbal manner with students. There exist different ways to animate and represent pedagogical agents, the most common being use of cartoon-style characters, playing real videos or employing 2D or 3D models.

Pedagogical agents can show affective reactions, have expressive behaviors, exhibit a personality, speak and listen to the user and also can offer a further dimension to the experience of human-computer interaction applications for learning purposes (Johnson and Lester, 2018). Their construction makes it possible to extend the experience within the learning environment by enhancing communicative setting, delivering a capability where the spoken output can be enhanced by expressive talking agents that can understand student speech, and are capable to speak to them. The use of pedagogical agents has led to significant progress in the creation and innovation of new learning technologies. These types of agents allow visualization of a better performance in the students and a remarkable improvement in the digital tools used in teaching-learning processes. How to design an animated virtual pedagogical agent which behaves much like a sensitive and effective human tutor is a very challenging task to carry out.

There are several motivations for using a virtual tutor in the form of an animated presentation agent in the teaching-learning process. These are: add expressive power to a system's presentation skills, help the students to perform procedural tasks by demonstrating them, serve as a guide through the elements of the scenario and, engage students without distracting or distancing them from the learning experience. By using robust artificial personalities as tutoring agents, it is possible to study the implementation constraints, effectiveness, and appeal of social interaction between tutoring agents and students. The intelligent software-agent technology has been suggested as a promising approach to extend ITSs in such way that the need for social context in learning can be fulfilled Ashoori et al. (2009). During the last decade, converging evidence indicates that learning gains can be achieved by designing computer programs that use pedagogical agents to foster the social agency. Research has shown that learning programs with well-designed animated pedagogical agents engage and motivate students, produce greater reported satisfaction and enjoyment by students, and produce greater learning benefits than programs without these agents (Yan and Agada, 2010).

Hautala et al. (2018) mention that evidence from national and international assessments indicates that a significant proportion of students in the world fail to meet grade-level academic standards in science. Previous research has indicated that virtual tutors can provide high-quality, individualized, and highly engaging science teaching equivalent to human tutoring. Results among children have also been encouraging and reviews indicate that having a visually present animated tutor does not seem to produce any learning gains over having animated tutors that are not visually present. However, this hypothesis has not been tested in young children. In a preliminary assessment of young children's impressions of studying with animated tutors, kindergarten and first-grade students reported that for them a virtual tutor is smart, cares about them, and acts like a real teacher who helps them to learn. Recent studies have suggested great potential for well-designed virtual tutoring systems to engage and motivate primary school students to learn science and to achieve learning gains comparable to those from human tutoring. The applicability of virtual tutors is yet to be studied in children, who have just started on their school path and are acquiring the basic scholastic skills, such as reading and comprehension in content areas, such as science.

The point of departure in the approach described in this chapter is that this agent technology can be used in the development of an effective pedagogical agent in the form of an intelligent virtual tutor for supporting and assisting the teaching-learning process in the first three grades of primary education in Mexico. Given the above, this chapter presents Mexican Intelligent Pedagogical Agent for Schoolchildren (MeIPeAS) as a prototype of a virtual tutor in the form of a new tool based on human-computer interaction techniques through the approach of serious games (Michael and Chen, 2005; Wouters et al., 2013) and educational software (Passey et al., 2018; Huang et al., 2019). Thus students in the first three grades of primary education in Mexico can interact with it while seeking to achieve understanding and master the knowledge acquired in each subject of each block included in all subjects to be studied.

The interactive virtual tutor here is intended to support the teaching-learning process within the classroom, not as a substitute for a real tutor or teacher, but as a pedagogical support element. This virtual tutor represents a new working tool that can be used in teaching, and helps in the increase and remarkable improvement of the educational quality from primary education, being one of the first pedagogical agents developed in

Mexico. This agent inhabits serious game acting as an individualized teaching-learning platform. With the use of the virtual tutor (Fig. 1), it will be verified that it is feasible to provide a supporting tool for the improvement of educational didactics, as well as a reinforcement of the subjects taught in the first three grades of basic education in Mexico. This demonstrates that means of reinforcement exercises for the subjects included in the current curriculum, a simple, creative, dynamic, intuitive and fun interaction mechanism is achieved, which assists and supports the teaching-learning process by impacting the user experience in a positive manner.

According to (Mayer, 2005), the cognitive theory of multimedia learning states that the narrated multimedia presentations help learners construct rich multimodal mental representations that lead to a deep learning of concepts. A large body of research indicates that relative to other presentation modes, such as texts with pictures, well-designed narrated multimedia presentations, in which a spoken voice explains concepts presented in illustrations or animations, optimize both the short-term retention of information and the transfer of learning to new tasks. Additionally Hautala et al. (2018) say that spoken explanations combined with visual illustrations, simple presentations in small steps, and control over the pacing of lesson content improve learning in multimedia settings. These design principles minimize the cognitive resources required for using the application, allowing users to maximize their focus on acquiring content knowledge. Early-grade students are only beginning to develop their basic scholastic and cognitive skills, such as reading and listening comprehension.

It is known that virtual tutors have a wide range of functionalities, which have been little exploited and applied in the educational field at the primary level. However, these functionalities allow offering mechanisms of interaction with students through an interactive dialogue by using text to speech, and even more sophisticated, the recognition and understanding of natural language or speech. In this chapter, a personalized virtual tutor for the primary education scenario in Mexico is presented. This virtual tutor, called 'MeIPeAS' was created for use as a pedagogical support mechanism, offering a unique attraction for current and future generations of schoolchildren. The virtual tutor has been validated in practice in public primary schools of the municipalities in the state of Mexico in Mexico. This validation is to analyze the impact of the user experience and the results of the research carrying relevant information on the reinforcement of topics taught in the classroom.

Fig. 1: Individual use of the pedagogical virtual tutor MeIPeAS within a teaching-learning environment for Mexican schoolchildren.

This chapter is organized as follows: Section 2 is devoted to explaining previous and recent efforts on pedagogical agents as virtual tutors and the use of chatbots as a class of conversational software agents. This is to examine the effects and benefits on students' learning. The main aspects of designing a pedagogical agent as a virtual tutor are discussed in Section 3. In Section 4, MeIPeAS is presented, a prototype of a virtual tutor for students in the first three grades of primary education in Mexico. The final remarks, conclusions, and future work are discussed in Section 5. Finally, Section 6 presents some future trends related to pedagogical agents as virtual tutors.

2. Pedagogical Agents Playing the Role of Virtual Tutors

Animated agents are graphical representations of characters used in computer interfaces where a human-computer interaction in a more personalized way and natural style is preferred to be present between users and agents. These virtual characters have been created for a wide variety of computer applications, such as virtual presenters, training agents, and pedagogical agents. In pedagogical applications, users tend to prefer a learning system which employs an animated agent to a similar one without an agent. Animated pedagogical agents are often designed as human-like characters that facilitate the learning process and support students in solving activities in learning platforms. Such agents are a useful aid for teachers in schools during learning activities. The use of animated agents in pedagogical applications has been effective in tutoring systems in which students can improve the learning experience by engaging in a dialogue or conversation with agents, that is, a learning environment can be enhanced with the addition of animated agents. When a pedagogical agent is facially animated in a learning platform, students experience a more natural learning process because the agent positively affects students' perception of their own learning experiences by its presence.

As shown by Atkinson et al. (2005), some of the motivation theories are used as part of the pedagogical agent design. The most frequently used one in pedagogical agent research is the social agency theory as a theoretical framework, assuming that sounds and images in pedagogical agents build social cues and trigger social responses, improving learning in student activities. As a final argument, it is argued that a pedagogical agent should be a teachable agent able to learn while teaching. Blair et al. (2007) defined a teachable agent as a one that works by letting students teach them and then assess their knowledge by providing a series of questions to solve problems.

Previous efforts on pedagogical agents include empirical researches about their uses, effectiveness, or limitations. The most representative examples of successful or agent-based learning applications are summarized by Ashoori et al. (2009). Some of these pedagogical applications are:

- *Hernan the Bug*: The first agent developed to inhabit virtual environments to help middle school students to understand botanical anatomy and physiology.
- *Soar Training Expert for Virtual Environments (STEVE)*: This was designed to interact with students in networked virtual environments and applied to naval training tasks.
- *Cosmo*: A life-like animated agent with deictic believability that occupies the Internet Adviser for helping learners on Internet packet-routing.

- *PPP Persona*: It was created to guide the learner through web-based material by using presentation acts to draw attention to elements of the web pages, and provide commentary via synthesized speech.
- *JACK*: A virtual human that exhibits a variety of different deictic gestures as a virtual presenter, who can point at individual elements on his visual aid and integrate these gestures into his presentation, moving over to the target object before his speech reaches the need for the deictic gesture.
- *ADELE*: An agent designed to operate over the web and able to extend the pedagogical capabilities of STEVE agent in a wider range of educational problems.
- *VINCENT*: A tutor-friend for on-the-job training to assist the student, promote his confidence and motivate learning.

Recent efforts on pedagogical agents show how agents and applications evolved to provide better learning platforms and solutions with fewer restrictions where students expect the agent to be believable as virtual mentors, be entertaining, easy to communicate, helpful, and diversified. A good example is described by Wise et al. (2005), where a virtual tutor called Ms. Readwrite gives hints and explanations to help children to figure out answers while learning about phonological awareness, reading, spelling, and comprehension. This virtual tutor helps children learn to recognize words, to read fluently, and to understand what they read, providing engaging practice with individualized support and focused hints.

Ashoori et al. (2009) also explained the most recent efforts on relevant pedagogical agents, the most relevant being the next ones:

- *Automated Lab Instructor (ALI)*: This agent provides flexible guidance to students interacting with virtual labs and was applied to chemistry and biology simulations. Its application to Virtual Factory Teaching System after that provided further evidence of its generality.
- *MASCARET*: A pedagogical multi-agent system for training proposed to organize the interactions between agents and to provide them reactive, cognitive, and social abilities to simulate the physical and social environment.
- *SKIP*: A puppet agent used to improve science learning through the use of digital puppets in peer teaching and collaborative learning settings.
- *Mediating Agent of MACES*: An animated agent that has the role of providing emotional support and promoting in the student a positive mood.
- *MENTOR*: A virtual teacher who monitors student activities and collaborates with other guide agents of the environment. It allows users to travel to virtual places to perform educational activities, talk with other users and mentors, and build virtual personae.

My Science Tutor (MyST) is described by Ward et al. (2013) to represent an intelligent tutoring system designed to improve science learning by students in 3rd, 4th, and 5th grades through conversational dialogues with a virtual science tutor. Individual students engage in spoken dialogues with the virtual tutor Marni, during 15- to 20-minute sessions following classroom science investigations to discuss and extend concepts embedded in the investigations. The spoken dialogues in MyST are designed to scaffold learning by presenting open-ended questions accompanied by illustrations or animations related to classroom investigations and the science concepts being learned.

A study, conducted with Turkish secondary school students to identify the impact of pedagogical agents on learners' academic success and motivation, was carried out by Dincer and Doganay (2015). In this study, four groups were formed. The first group received education via fixed pedagogical agents, the second group had the option to choose among several pedagogical agents, the third group received the education without pedagogical agents and finally, the last group received the same education through traditional (non-computer) way. The study was conducted during a four-week program introduced to students via Microsoft Excel and the data were gathered by pre- and post-test method. The findings revealed that interfaces impacted motivation and accordingly academic success in a positive way. As a result of the study, it is suggested that learners should be provided programs that can be personalized, depending on learners' needs and preferences. Four pedagogical agents were employed (Tuna, Ada, Ali & Zipzip) (Fig. 2).

Although several studies have been conducted to examine the effects of pedagogical agents on learning, little is known about gender stereotypes of agents and how those stereotypes influence student learning and attitudes. One aspect of pedagogical agent implementation that deserves thorough investigation is the effect of the agent's physical appearance. Researchers have found that learners stereotype pedagogical agents by their physical appearance and non-verbal cues. Gender stereotypes are unconscious thought processes that guide expectations on how each gender should look, speak, and behave. A study done by (Schroeder and Adesope, 2015) investigated if the pedagogical agent's gender influences cognitive and affective outcomes in learner-attenuated system-paced environments. This study investigated two research questions to explore how a pedagogical agent's gender might affect cognitive and affective outcomes in pre-service teachers' learning with a pedagogical agent in a learner-attenuated system-paced learning environment. How does learning with a male pedagogical agent affect the learner's free recall, multiple-choice, and transfer scores compared to learning with a female agent? How does learning with a male pedagogical agent affect the learner's perception of the agent compared to a female agent? The results revealed that the gender of the agent did not produce any statistically significant effects. This indicates that the effects of learning with pedagogical agents may be independent of the agent's gender. The male and female agents are shown in Fig. 3.

Kim and Baylor (2016) found that what students wanted from a pedagogical agent was a good teaching ability to be knowledgeable and the ability to give motivation and being friendly and kind when using various instructional strategies in an interactive learning environment. Students significantly learn better and have greater motivation

Fig. 2: Tuna, Ada, Ali, and Zipzip as instructional agents to teach Microsoft Excel to Turkish secondary school students (*Source*: Dincer and Doganay, 2015).

Fig. 3: Male and female pedagogical agents used to investigate the impacts of agent gender in an accessible learning environment (*Source*: Schroeder and Adesope, 2015).

when working with two agents (split persona effect) than with one mentor agent. Mixed design factors of pedagogical agent research significantly improve learning performance and student behavior. According to Johnson and Lester (2016), due to pedagogical agents becoming more common and complex, it is important to develop a better understanding of how best to combine interactive agents with teachers and students, to complement and make the learning process most effective and positive. Pedagogical agents should not be seen as taking the place of teachers but as a supporting tool in the learning process in the classroom (Taub et al., 2016). An on-going work to endow a virtual agent tutor with the capability to provide and interpret feedback in a pedagogical interaction with learners is presented by Taoum et al. (2016). The objective of the tutor's behavior is to deduce, according to learner's feedback, at which stage the learner is and which exact cognitive process occurs. The result of this inference is stored in the learner's model. The tutor behavior is defined according to who (tutor or learner) takes the initiative to interact. Most of the time, the tutor agent can be the initiator of the interaction. This tutor is shown in Fig. 4. In the picture on the left, the virtual tutor informs about the first action to perform; in the picture on the right, the learner asks for more information about the object to manipulate (neoplastine). Through its communication behavior, the virtual agent interprets this question and, using its knowledge about the environment, describes the object.

Song (2017) mentions that learning-by-teaching pedagogy is one of the more effective approaches to learning. In a face-to-face learning-by-teaching situation, the role of the learners is to teach their peers or instructors. In virtual environments, learners play an active role by teaching a computer agent, named as the teachable agent. In this

Fig. 4: A virtual tutor explaining the way to manipulate and object for a blood analysis procedure (*Source*: Taoum et al., 2016).

way, a virtual learning-by-teaching environment is presented. A communication method was adopted in this system which facilitates the interaction between the learner and the computer agent, specifically in K-12 students' mathematics learning. According to Abdullah et al. (2017), discovery-based learning is an educational activity paradigm through well-specified experiences that are designed to foster particular insights relevant to curricular objectives. It differs from many of the applications in which pedagogical agents have traditionally been used. In that the knowledge desired for the student is never explicitly stated in the experience; rather, the child discovers it on her or his own. Testing also differs, as the goal is a deeper conceptual understanding, not easily measured by right or wrong answers. In this way, a pedagogical agent designed to support students in an embodied, discovery-based learning environment was created. Discovery-based learning guides students through a set of activities designed to foster particular insights. In this case, Maria, an animated agent, explains how to use the Mathematical Imagery Trainer for Proportionality, provides performance feedback, leads students to have different experiences, and provides remedial instruction when required. This agent can be seen in Fig. 5, which was programmed to lead students through a series of activities and can execute a sequence of action blocks. Each block consists of a subset of spoken audio, facial animation, lip-syncing and body animation, including arm and finger gestures.

Based on the fact that attention-deficit/hyperactivity disorder (ADHD) is a persistent condition associated with impairment in educational functioning, professional position, and social relationships, which is determined by three basic symptoms: inattention, hyperactivity, impulsivity, drawing the attention is the first key step to enhance learning as is shown by Mohammadhasani et al. (2018). An environment embedded with a pedagogical agent in computer-assisted instruction has been designed to support learning through gaining and guiding attention to relevant information for these students. The study is used to investigate how much the presence of Koosha as a virtual tutor and motivator can improve learning in ADHD students. The agent is shown in Fig. 6. This study employed a pretest and post-test experimental design with a control group. The statistical population was 30 boy students with ADHD in primary school from the north of Iran. The participants were randomly assigned to work with either an agent presenting

Fig. 5: Maria, a virtual tutor, is able to explain how to use the Mathematical Imagery Trainer for Proportionality (*Source*: Abdullah et al., 2017).

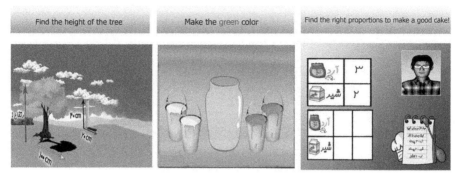

Fig. 6: Koosha, a virtual tutor and motivator in math learning for ADHD students (*Source*: Mohammadhasani et al., 2018).

a multimedia program or without an agent in mathematics. The results suggested that experimental and control groups showed a significant difference in mathematics achievement. According to this research, using the pedagogical agent can enhance the learning of ADHD students; so it can be considered as a valid school-based intervention for these students.

Hauatala et al. (2018) present a virtual tutoring system that teaches first-grade students in Finland new ideas and principles to explain various animal behaviors, structures, and functions, employing short audio-visual lessons interleaved by multiple-choice problems with guided feedback. The lessons are organized in learning progressions, starting with an intriguing question, then teaching required concepts one by one with exercises, and finally practicing the aim of the lesson. The design of the virtual tutoring system, Mindstars Books (MSB), integrates ideas from the cognitive theory of multimedia learning, formative assessments, and the Dual Situated Learning Model. While these theories provide a general framework for the design of the presentation format, instructional interaction, and science content progressions, respectively, special attention is paid on how to design the system to optimize learning in little children.

Another way to see a virtual tutor is in the form of a chatbot. Chatbots are a class of conversational software agents. According to Khanna et al. (2015), any chatbot program understands one or more human languages by natural language processing. Due to this, the system interprets human language input using information fed to it. A chatbot may also perform some productive functions, like calculations, setting-up reminders or alarms, etc. A popular example of a chatbot is the Artificial Linguistic Internet Computer Entity (ALICE). Bot (Wallace, 2009), which utilizes Artificial Intelligence Markup Language (AIML) pattern-matching techniques, is an open, minimalist, stimulus-response language for creating bot personalities as open standards for chatbots. The use of conversational agents in virtual communities can contribute to better user interaction and assisted learning at higher education levels. An example is BOT-BLOG (Moreno and Toro, 2010), a conversational agent developed to solve doubts that students may have in the subject of Requirements Analysis that is taught as a second cycle subject in computer engineering career that is taught at the Ourense Campus of the University of Vigo. The chatbot is aimed at encouraging students to ask questions. Depending on asked question asked, it will present a more or less complete answer by using in an integrated way various technologies and techniques into the WEB 2.0. The typical processing steps of a chatbot are input cleaning by using removal and substitution of characters and words, like smileys and contractions by using a pattern matching algorithm to match input templates against the cleaned input, determining the response templates, and generating a response (Masche and Le, 2017). Another example is Mike (AlKhayat, 2017), which was used in a study to explore the effectiveness of using chatbots in promoting English language proficiency. The study focused on 22 undergraduate students divided across three distinct groups at an Arab university in Saudi Arabia (Fig. 7). The study investigates the students' ability to practice English by chatting with a virtual agent. The researcher used situational learning and constructivist learning theories as theoretical underpinnings. Survey results of students' perceptions and teachers' reflections on their teaching strategies show that students' levels and language skills are important factors in determining the effectiveness of chatbot used in the classroom.

In order for a chatbot to construct a supposedly intelligent answer, the user's input, which is in the form of a written text or sound signals, is analyzed morphologically, syntactically, semantically, pragmatically, and contextually in order to generate an answer that abides by the same linguistic rules and relates to the same discourse (AlKhayat, 2017). The Turing Test is one of the most popular measures of intelligence of such systems. Bachle et al. (2018) defined a chatbot system as a software program that possesses some kind of intelligence or at least a set of algorithms that allows interaction between humans and computer systems by using natural language. The language might be written text or spoken language. In some cases, they are designed to have an ongoing conversation; in other cases, they may be commanded to execute tasks.

A list of various chatbots can be found at (Chatbots.org, n.d.). While some are designed to compete in several different chatbot competitions, like the Loebner Prize (n.d.), others are used as a tool for entertainment or information retrieval. At this moment, Mitsuku (Worswick, n.d.), an artificial intelligent lifeform living on the net, a four-time winner of the Loebner Prize Turing Test (2013, 2016, 2017 and 2018 editions), is the world's best conversational chatbot. Mitsuku uses AIML to understand and respond to people. Mitsuku was created through the ALICE's AIML files and includes the ability

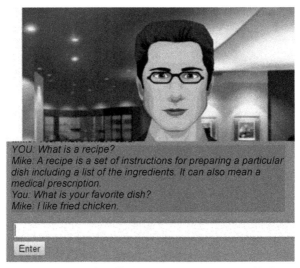

Fig. 7: Mike, a chatbot to explore the effectiveness in promoting English language proficiency in undergraduate students at an Arab university in Saudi Arabia (*Source*: AlKhayat, 2017).

to reason with specific objects. Users can talk to Mitsuku on Mousebreaker Games, Facebook Messenger, Twitch group chat, Telegram and Kik Messenger under the username Pandorabots. For a while Mitsuku was available on Skype, but was removed by its creator Steve Worswick, who has written several AIML files to add extra content to her. If somebody has any AIML bots, is free to incorporate any of these files into proper bots. These files are designed ideally for use with Pandorabots, but should work in other AIML interpreters and should be easily convertible into other chatbot languages. Mitsuku has different *avatar* representations, some of which are shown in Fig. 8.

In this way, the chatbot technology is a viable technology for certain types of ambient-assistive technology. However, it comes with some limitations and is not suitable for all target groups. Further research might investigate possible useful combinations with other interaction technologies (e.g., speech recognition, gesture control, etc.). Therefore, the use of chatbots in workplace environments is increasingly being investigated. As it is mentioned by AlKhayat (2017), chatbots are useful internet applications, which become more useful when applied in the classroom to enhance the learning experience. There are still some drawbacks that hinder effective interaction with the chatbot (considering the fact that it is not human). Consequently, more studies need to be conducted to explore other applications and methodologies for using chatbots that could be useful to the students.

In 2007, Eviebot (Existor Ltd., 2016), a female embodied chatbot with realistic facial expressions, went online. Evie is a learning artificial intelligence and an advanced, emotional chatbot *avatar*. The Evie chatbot has had a huge impact on social media over the last few years. She can speak several languages and is probably the most popular artificial personality on YouTube. She has appeared in several videos by PewdiePie, the most subscribed YouTuber in the world and other YouTube channels. The things she says were learned from a human being at some point in the last 10 years. The information is stored in a database which Evie looks through every time she needs to say something. Evie is shown in Fig. 9 on the left side. Additionally, Pewdiebot (Existor Ltd., 2018), a male counterart for Eviebot, which is an *avatar* representation for the world's most subscribed YouTuber Pewdiepie (Fig. 9 on right side).

Fig. 8: Mitsuku's *avatars* (Left): The most common representation as a 2D *avatar* (Worswick, n.d.) (Right): The 3D Twitch *avatar* (*Source*: Pandorarobots, n.d.).

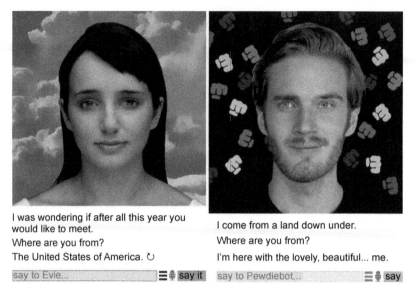

I was wondering if after all this year you would like to meet.
Where are you from?
The United States of America. ↻

I come from a land down under.
Where are you from?
I'm here with the lovely, beautiful... me.

say to Evie...   ☰ 🎤 say it   say to Pewdiebot...   ☰ 🎤 say

Fig. 9: (Left): Evie, an emotional chatbot, able to learn from people (*Source*: (Right): Pewdiebot, an *avatar* representation for the world's most subscribed YouTuber Pewdiepie (*Source*: Existor Ltd., 2018).

Pewdiebot's artificial intelligence allows it to have more foolish things to say for itself. Both chatbots share the same technology with Cleverbot (Carperter, 2019) and can speak several languages. Unlike some other chatbots, Cleverbot's responses are not pre-programmed. Instead, it learns from human input. From the humans' type, the chatbot system finds all keywords or an exact phrase, matching the input. After searching through its saved conversations, it responds to the input by finding how a human responded to that input when it was asked, in part or in full, by Cleverbot.

3. Designing a Pedagogical Agent as a Virtual Tutor

Johnson et al. (2000), as cited by Morton and Jack (2005), detail the types of interaction that agents can display which benefit the learning context, and also describe the following benefits that animated pedagogical agents can bring to human-computer interaction applications:

- Pedagogical agents can teach students how to perform a particular task. They can virtually show how to complete a task into a virtual learning environment.

- Pedagogical agents can act as navigational guides to students in complex learning environments.

- Pedagogical agents can steer a student's attention to something in the virtual world by way of gaze behaviors and deictic gestures.

- Pedagogical agents can provide non-verbal feedback to a student's input or actions, as well as verbal feedback.

- Pedagogical agents can express conversational signals which people are accustomed to in human-human interaction, for turn-taking, expressing personal opinions, or acknowledging the user's utterance.

- Pedagogical agents can convey emotion to the user which, in turn, may elicit emotion from the user and serve to increase learning motivation.
- Pedagogical agents can serve the role of a virtual teammate and working in a team is an element of the task design. Here, the agent can either act as an instructor, helping the student to accomplish a task, or substitute as a missing team member, allowing the student to practice working in a team.

According to Mitrovic and Surweera (2000), an agent can exist within the learning environment, but it is also possible for an agent to exist in a separate window. In addition, it is preferred that the behavior be generated ideally online and in a dynamic way, instead of being created offline or manually. Dynamic behavior is better and corresponds to the changes in the learning environment. An animated pedagogical agent can have a positive outcome on the learning experience by personification whereby the student tries to impress the agent; the agent collaborates with the student, and provides several types of feedback including causal, congratulatory, deleterious, assistive, background, and motivational responses; and by engagement (Yan and Agada, 2010).

Dincer and Doganay (2015) reviewed the classifications for pedagogical agents and mentioned that these are generally considered in two main groups—design/presentation forms, and tasks/functions while classified. Regarding their designs/presentations, pedagogical agents are classified in different forms, such as visuals, audible, and text-based. In the case of visual pedagogical agents, these are in the form of one of the following representations:

- *Human-like*: A real human image or animating a real human image by drawing.
- *Cartoon film character*: Animating a cartoon film character or a 2D shape/figure.
- *Gestures*: Using human gesture images or drawings.

Audible pedagogical agents only include the guidance of a person by talking in the background. These agents can process and understand human speech and talk to the users by text to speech mechanism. On the other hand, textual pedagogical agents involve guiding users by providing sentences or words and also processing entered text by the users. That is why they are able to establish a one-way or two-way dialogue or communication with users.

Regarding their tasks/functions, pedagogical agents have been mainly classified as follows:

- *Smarts*: Agents, which can learn by using Artificial Intelligence mechanisms and respond to users naturally and credibly.
- *Guides*: Agents, which inform users about the use of software or train users to use physical tools.
- *Subsidiaries*: Agents which provide clues to users about the topic and questions.
- Other types of tasks or functions are meant to be informational or informative, raters, advisors, and experts.

Mohanty (2016) argued that an intelligent tutoring system exhibits characteristic features similar to a human tutor, such as being able to answer student questions, detect misconceptions and provide accurate feedback. In this manner, the agents are also seen as having social (and relational) artifacts, besides cognitive assistance. Thus, in e-learning environments, the pedagogical agents play multiple roles, such as tutors, coaches, and actors, experts, motivators and mentors, learning companions, change agents, and lifelong learning partners. Additionally, learners may stereotype agents according to their appearance. An appealing visual appearance enables agents to function as active social role models for e-learners. Thus, their defining roles are:

- *Adapt*: A pedagogical agent evaluates the learner's understanding throughout the interaction, just as a human teacher would, and adapts the lesson plan accordingly. Agents will not move on to more sophisticated concepts until it is clear that the learner has a good understanding of the basics; if learners continue to have difficulty, the agent can provide additional instruction.

- *Motivate*: Pedagogical agents can prompt students to interact by asking questions, offering encouragement and giving feedback. They present relevant information, offer memorable examples, interpret student responses, and even tell a clever joke or two.

- *Engage*: Pedagogical agents have colorful personalities, interesting life histories, and specific areas of expertise. They can be designed to be the coolest teachers in school.

- *Evolve*: Pedagogical agents can be revised and updated as frequently as desired to keep learners current in a rapidly accelerating culture. They can search out the best or most current content available on the web to enrich the lessons that someone else has previously designed.

Virtual tutors are viewed as conversational systems. These systems are chatbots and dialogue systems; virtual tutors are a mixture of both systems. According to Masche and Le (2017), these two categories of conversational systems are not clearly defined, but the main differences in the architecture between dialogue systems and chatbots are the natural language-understanding component and the dialogue manager. A dialogue system denotes a system, which is able to hold a conversation with another agent or with a human. Dialogue systems make use of more theoretically-motivated techniques and often are developed for a specific domain, whereas chatbots are aimed at open-domain conversation. While a typical chatbot is built, based on a knowledge base, which comprises a fixed set of input-response templates and a pattern-matching algorithm, a dialogue system typically requires four components: a pre-processing component, a natural language-understanding component, a dialogue manager, and a response generation component (Lester et al., 2004).

In order to build a virtual tutor as a chatbot, the following languages can be used (Masche and Le, 2017):

- *Pattern matching*: To generate an appropriate response to the user's utterance by analyzing the user's input and matching it to the keywords in a pre-specified dictionary. For a found keyword, it applies an associated input-response rule.

- *Cleverscript*: The main concept of Cleverscript is based on spreadsheets. Words and phrases that can be recognized (input) or generated (output) by Cleverscript are written on separate lines of the spreadsheet. Cleverscript and the concept of this chatbot authoring language make the development of chatbots relatively easy.

- *Chatscript*: Similar to Cleverscript, Chatscript is based on pattern matching. Another special feature of Chatscript is the so-called Concept Set, which covers semantic-related concepts of a constituent in user input.

- *AIML*: An XML-based language for developing chatbots, AIML has established itself as one of the most used technologies in today's chatbots. AIML is based on pattern matching. An AIML script consists of several categories, which are defined by the tag. Each category consists of only one tag, which defines a possible user input, and at least one tag, which specifies the chatbot's response to the user's input. Like Cleverscript, AIML makes use of wildcards in order to cover a large possibility of user's inputs. In order to interpret these AIML tags, a chatbot needs an AIML interpreter. Various AIML interpreters, using different programming languages, such as Java or Python, are available. Since developing AIML chatbots does not require skills in a specific programming language, this technology facilitates the development of chatbots.

- *Language tricks*: Different language tricks are usually used by chatbots, including canned responses, a model of personal history, no logical conclusion, typing errors, and simulating keystrokes. Canned responses are used by chatbots in order to cover questions/answers of the user that are not anticipated in the knowledge base of the chatbot. A model of personal history enriches the social background of a chatbot and pretends the user to be a real person. Statements with no logical conclusion are embedded in chatbots in order to enrich small talks. Typing errors and simulating keystrokes are usually used to simulate a human being, who is typing and making typo errors.

The typical components of a dialogue system are the next ones (Masche and Le, 2017):

- *Preprocessing*: Most dialogue systems process the user's input before it is forwarded to the Natural Language Understanding component. The tasks of pre-processing are divers. Dialogue systems mostly deploy the following natural language preprocessing tasks: sentence detection, co-resolution, tokenization, lemmatization, POS-tagging, dependency parsing, named entity recognition, semantic role labeling.

- *Natural language understanding*: To understand the input made in the form of sentences in text or speech format.

- *Dialog manager*: The dialogue manager is responsible for coordinating the flow of the conversation in a dialogue system. Approaches to developing a dialogue manager are categorized in one of the following: finite state-based systems, frame-based systems, and agent-based systems. In finite state-based dialog systems, the flow of the dialogue is specified through a set of dialogue states with transitions denoting various alternative paths through a dialogue graph. Frame-based systems ask the user questions that enable the system to fill slots in a template in order to perform a task, such as providing train timetable information. The approach underlying agent-based dialogue systems is detecting the plans, beliefs, and desires of the users and modeling this information in a Belief-Desire-Intention agent.

- *Response generation*: Adding different types of the same expression to an utterance and generating a smooth response is based upon a set of template sentences. The generation of utterances is based on the last estimated dialogue-act, which consists of templates containing text, pointers, variables, and other control functions.

- *Special features*: In order to make conversational systems more like humans. For instance, some systems are able to learn from conversations and can apply this knowledge later. Multimodal systems can communicate with the user through both text and speech channels. With the development of embodied conversational agents, features like gestures, facial expressions or eye gazes become increasingly important. Developers of pedagogical agents often include graphics, videos, animations, and interactive simulations into their system to increase the student's motivation.

While chatbots deploy dominantly pattern-matching techniques and language tricks, most dialogue systems exploit natural language technologies. Most chatbots participated in the Turing Test contests, like the LoebnerPrize, while dialogue systems are mostly evaluated by the pre-/post-test, quantitative, or qualitative methods. This can be explained by the fact that dialogue systems are more goal-oriented to improve learning gains of students while chatbots serve small talks in different domains. The tendency of applied technologies for conversational systems is becoming more Artificial Intelligence-oriented and deploying more natural language-processing technologies (Masche and Le, 2017).

4. Mexican Intelligent Pedagogical Agent for Schoolchildren (MeIPeAS)

In the initial stage, there are some aspects to keep in mind while developing the virtual tutor MeIPeAS within the Mexican education scene, which are as follows:

- The benefited primary schools are some of those belonging to the municipalities of the state of Mexico in Mexico.
- The topics of the courses taken within the first three grades of primary education will be considered according to the current curriculum, which consists of Spanish, Mathematics, Exploration of Nature and Society, Civic and Ethical Education, and finally, Artistic Education.
- Only some reinforcement exercises will be created, which will be taken as models to create the remaining ones in a final version of the virtual tutor.
- Students can choose the gender of the virtual tutor, as they feel better.
- The voice of the virtual tutor will be synthetic and will be reflected through a bidirectional dynamic dialogue, both to the student through text to speech and to the tutor through speech recognition.
- The virtual tutor will not be able to express emotions either verbally or visually. However, for a final future version, this aspect will be fully covered.

4.1 Current Educational Landscape

From the didactical point of view, there are numerous approaches to learning, such as learning by observation, learning by inquiry and investigation, learning by doing, individually, face-to-face and in groups, experimental learning, learning by evaluation and reflection (Pivec et al., 2004).

In Mexico, a good quality education (Cuéllar, 2012) can be sustained if it contains the following attributes: relevance, pertinence, internal efficacy, external efficacy, quality, sufficiency, good educational results, and equity. However, within education institutions at basic levels, despite the creation and implementation of new methodological teaching proposals towards mastery of competences, in practice there have been disadvantages, such as numerous groups, which hinder a good start-up of an adequate teaching-learning process.

An integral reform in education in Mexico (Guichard, 2005) at each of the levels that integrate basic education, must be focused on providing students with a learning scheme based on development and application of competencies. This learning scheme has lagged due to the educational models and teaching-learning media that are applied (Treviño Ronzón and Cruz Vadillo, 2014), where, in general, the theory is reviewed and explained by using media and materials, such as printed texts or digital or virtual resources. Students understand little information and acquire knowledge that they do not know about how to put it into practice beyond the classroom. In this way, these resources have been very badly exploited, leading to failure in inculcating the base for the development of basic science training and acquisition of notions about information technology and communications.

Trying to remedy the problem above mentioned and as a point in favor, it is well known that the new information and communication technologies have been little used (incredible as it may seem), being partially or totally outside the educational processes, since they try to improve the educational didactics. In addition, these technologies facilitate the achievement of a broad and better integral development of the students, as much in their abilities to learn as in the application of knowledge at basic levels of education.

In Mexico, the use of computing applications within the educational didactic is very little applied or failing that, it is not exploited correctly in several of the school levels, specifically in primary education. When developing computing applications in the educational field, it is intended that their use provides aid in teaching in the classroom, or be a useful tool to assist learning at home. This fact leads to the use of the term, educational software, which is based on learning models, as well as on concepts of cognitive and constructivist psychology. On the other hand, there is a subcategory known as 'virtual educational tutors', on which this paper is based.

The last integral reform in basic education in Mexico faces the main challenge to raise the quality of education in the country, through continuous improvement of technological resources and high training of teachers and teaching to students, so that both take advantage of the innovation present in the use of new didactic resources in the classroom. This may be possible through the incorporation of information and communication technologies in professional training and pedagogical processes (Pelgrum, 2001), taking, for example, the following:

- The use of digital tools and resources to support the understanding of knowledge and concepts.
- To plan and manage research, using information and communication technologies.
- To use models and simulations to explore some topics.
- To generate original products with the use of information and communication technologies, in which critical thinking, creativity or the solution of problems based on real-life situations are used.
- To make responsible use of software and hardware, whether working individually, in pairs or as a team.
- To inculcate ethical, safe and responsible use of the Internet and digital tools as well.

Within the plans and programs of study of basic education in Mexico, the consolidation of a curriculum that encourages the development of competence in the use of information and communication technologies, is in response to the social demand and the quality of the Mexican public school (Scott et al., 2018). To achieve this, it is known that improvement in the infrastructure and equipment of the primary schools, in particular of their laboratories and workshops, is a priority. To do this, it must be based on digital technologies to equip students and teachers with devices and technological links.

It should be noted that in the first three grades of basic education, it is where the bases for the development of basic science training and the acquisition of notions about technology are established as mentioned by Boudourides (2003). The context is clear—any educational reform should not evade digital skills standards and through these, it will be possible to teach the proper use of information and communication technologies, which is the fundamental basis for the development and acquisition of skills (Abbot, 2003; Bhattacharjee and Deb, 2016).

4.2 MeIPeAS' Implementation within an Intelligent Tutoring System and Virtual Learning Environment

In order to personify the virtual tutor MeIPeAS (Fig. 10), GIF images of a cartooned aspect or appearance were used, as well as the corresponding animation to represent the speech action according to the case of the virtual tutor genre, male (Valentin) or female (Valentina). Audio and voice animations are synchronized in each GIF image to denote a better naturalness during the speech interaction with the student. One of the highlights in the design of the virtual educational tutor MeIPeAS is the human-machine interaction, with attractive designs to capture the attention of students and achieving an effective natural relationship in real-time.

Different pedagogical issues were taken into consideration in order to develop the prototype (McLaren, 2015; Van Manen, 2016; Bascope et al., 2019). These issues were essential to offer a better virtual tutor to the target student population. It should be remembered that this population only covers the first three primary school grades in Mexico. With the aim of making the virtual tutor more lucid, fancy, attractive, easy to use,

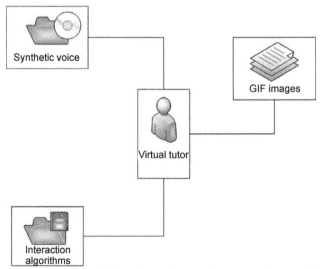

Fig. 10: Personification through synthetic voice, audio and GIF cartoon images of the virtual tutor, MeIPeAS.

funny and colorful for students, a window-based navigation scheme is used to implement the prototype here presented.

4.3 Module Description

One of the most outstanding aspects in the design of the educational virtual tutor MeIPeAS is the human-machine interaction, with attractive designs to capture the attention of the students, achieving an effective natural relationship in real-time. The general outline of the prototype is shown in Fig. 11 and which is composed of three interconnected modules and multimedia components. The functionalities of each module are briefly described as follows:

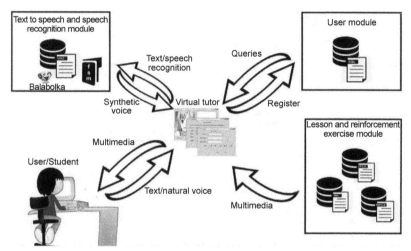

Fig. 11: General outline of the operation, design, and implementation of the educational virtual tutor, MeIPeAS.

4.3.1 User Module

This module has a database where the basic data of the users will be registered, which are their names, paternal surname, maternal surname, age, grade, gender, password that will automatically generate the system, the school to which they belong, as well as the municipality where they are located. Thus, the virtual tutor will have information about each student to interact in a personalized and unique relationship.

Figure 12 shows the main window when the virtual tutor is running, where students are asked to indicate their grade, gender, and type of virtual tutor (male or female). In Fig. 13, the log in window is shown. All students must be registered to be able to log in with the virtual tutor, MeIPeAS. The user registration window and required data to register can be seen in Fig. 14.

4.3.2 Lesson and Reinforcement Exercise Module

When a student has begun a session with the virtual tutor, MeIPeAS, the welcome window shown in Fig. 15 gets displayed. At each learning session, the student must choose a school subject and its respective block and topic to obtain a review lesson and information about

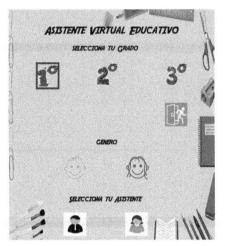

Fig. 12: Main window when executing the virtual tutor, MeIPeAS.

Fig. 13: Login window with the virtual tutor, MeIPeAS.

the selected subject. At the end of the lesson, the student has the possibility to accept the challenge of responding to a reinforcement exercise. Figure 16 shows the window used to start a reinforcement exercise. However, it is possible to cancel the exercise by exiting it and taking another one. If a student's answer is correct, MeIPeAS congratulates him/her, but when the answer is wrong, the student is motivated to try it again until the answer is the correct one. This motivation is to enhance the student's self-esteem and promote a higher comfort learning and practice to avoid him/her to abandon the activity by getting him/her to complete it successfully.

This module is formed by the first database of images that are used in the reinforcement exercises and the possible personifications of the virtual tutor (female or male), with a caricature style; and a second database with explanatory videos on the topics that are addressed in the curriculum of the first three grades of primary education in Mexico.

Fig. 14: Student registration window to log in with the virtual tutor, MeIPeAS.

Fig. 15: Student welcome window when logging in with the virtual tutor, MeIPeAS. (Left): A male representation, called Valentin is welcoming a boy student, called José. (Right): A female representation called Valentina is welcoming a girl student, called Laura.

Fig. 16: Menu of exercises according to each subject and its blocks and topics. Laura, a girl from the first degree is asked to select a reinforcement exercise.

4.3.3 *Text to Speech and Speech Recognition Module*

This module performs the action of converting the text introduced by the student through the keyboard or the options that he/she chooses in the personification of the virtual tutor's voice, that is, text-to-speech conversion. In addition, it is responsible for recognizing and processing the student's speech at the time the virtual tutor asks a question and waits for a verbal response from the student. As it is possible to notice, Spanish was used as the interface language given in the windows of the prototype as well as in the mechanism of interaction through the bidirectional dialogue with the virtual tutor. This is because in Mexico, Spanish is the official language and the vast majority of schoolchildren who study in the first three grades of primary education are having their first approaches with a second language, such as English as a foreign language. A database with XML files, based on the Java Speech API Markup Language (JSAML), is managed and whose structure is in the form of a finite state machine (FSM), which is considered for making decisions corresponding to the speech actions performed by students and the virtual tutor MeIPeAS (Fig. 17).

For the functioning of this module, authorship code and open source solutions are used as the command-line text to speech utility voice.exe from Eli Fulkerson and

```xml
1    <?xml version="1.0" encoding="UTF-8" standalone="no"?>
2    <fsm name="initialDialogue">
3      <state name="beginDialogue" type="begin" />
4      <state name="statement1" type="normal" />
5      <state name="endDialogue" type="end" />
6      <transition from="beginDialogue" to="endDialogue">
7        <conditions />
8        <actions>
9          <action type="query" emotional_appraisal="0.4">
10           <subaction>
11             periodOfDay
12           </subaction>
13         </action>
14         <action type="random" emotional_appraisal="0.7">
15           <subaction>
16             Qué bien [UserName]!
17           </subaction>
18           <subaction>
19             Magnifico [UserName]!
20           </subaction>
21           <subaction>
22             Estupendo [UserName]!
23           </subaction>
24         </action>
25         <action type="random" emotional_appraisal="0.6">
26           <subaction>
27             Gracias por utilizar esta herramienta de aprendizaje. Te veo pronto.
28           </subaction>
29           <subaction>
30             Espero tu pronto regreso al uso de esta herramienta de aprendizaje.
31           </subaction>
32           <subaction>
33             Te felicito por utilizar esta herramienta de aprendizaje. Nos vemos pronto.
34           </subaction>
35         </action>
36       </actions>
37     </transition>
38   </fsm>
```

Fig. 17: Example of an XML file in the form of a finite state machine (FSM). This file is to say goodbye to a student when exiting a learning session and finishing the interaction with MeIPeAS.

speech.exe from Balabolka. However, MeIPeAS was created to use commercial software, such as Microsoft Speech Platform, CereProc, and Amazon Speech Recognition Software.

At the moment, only students can register and log in with the virtual tutor. However, it is intended that the teacher can do it equally but in such a way that he can have the option to add and modify lessons. A more ambitious challenge would be that this prototype could be available via online on Internet, as well as be able to run on the various platforms, such as the Android, IOS, Windows, and Linux systems.

4.4 *Validation in a Real Scenario*

Some primary schools from the municipalities of Atizapan de Zaragoza, Tlalnepantla de Baz and Nicolas Romero from the state of Mexico in Mexico were pilot testing scenarios. The prototype was presented to teachers and students from the first grade of basic education, as shown in Fig. 6, doing the following:

- It was explained to the students and teachers about the prototype, the objectives that are seeking to be achieved, the benefits that this can have within the educational field and the way of interacting with the virtual tutor.

- A demonstration of the operation of the prototype was given, explaining how it is executed, the content of each menu, the way they should register, how to log in and use the exercises, as well as the interaction with the virtual tutor.

- Once the explanation was completed, teams of students were formed to interact with the virtual tutor prototype.

An example of a reinforcement exercise is given in Fig. 19. It is observed that the student must identify the animal shown by its name; in this case, the correct answer is a gorilla. Another example of reinforcement exercise is shown in Fig. 20, where a student has to identify the corresponding image of a body part asked by the virtual tutor, MeIPeAS. In both cases, MeIPeAS is representing the male character, Valentin. The last example is presented in Fig. 21, where the student needs to identify an asked number. In this case, MeIPeAS is representing the female character, Valentina.

At the end of the dynamics of interaction between students and teachers with the virtual tutor, 5 questions were asked on their appreciation of the prototype to collect their

Fig. 18: Gonzalo, an undergraduate student, who worked on the virtual tutor project is introducing MeIPeAS to schoolchildren.

Fig. 19: Reinforcement exercise related to a Spanish topic where MeIPeAS is asking a student to identify the animal shown by the central figure.

Fig. 20: Reinforcement exercise related to a Spanish topic where MeIPeAS is asking a student to identify the animal shown by the central figure.

user experience. The first three with yes/no, the fourth with two possible answers, fun and boring, and for the last, the answer was left open. These questions were as follows:

1. Do you know how to use a computer (basic level)?
2. Does your elementary school have a computer room?
3. Do you know what a virtual tutor is?
4. What did you think about the prototype of the virtual tutor?
5. How do you think a virtual tutor should be?

Fig. 21: Reinforcement exercise related to a Maths topic where MeIPeAS asks a student to identify and select the correct number according to its symbol and graphical representation.

In a total of 147 students, seven were five years old (5 per cent), 125 were six-years old (85 per cent) and 15 were seven-years old (10 per cent). Those who interacted with the virtual tutor comprised 72 boys (49 per cent) and 75 girls (51 per cent). Table 1 shows the percentages in the collected answers for the first four questions. It was mainly in the last question that students were unanimous in their answers when they said that a virtual tutor should be friendly, human-looking, colored and not scary. From the answers, it was verified that the user experience was positive and the proposed virtual tutor, MeIPeAS fulfilled its task as a support tool in the teaching-learning process. It was welcome and well appreciated by teachers and students.

Even the teachers were asked the same questions. In all there were six teachers with three males (50 per cent) and three females (50 per cent). Table 2 shows the percentages in the answers given by them to the first four questions. In the last question and in a similar manner as given to the students, the teachers agreed that a virtual tutor must not only fulfill its pedagogical purpose and be attached to a teaching-learning method adopted in schools, but also serve as a support and complementary tool within the classroom.

It is worth mentioning that there was no notable preference for choosing a gender representation over the other. However, a female character can induce better confidence in students because she is quite closely related to a maternal figure who gives protection and comprehension to schoolchildren. A male character can be seen as a respectful and strict figure that imposes order and discipline among schoolchildren.

Table 1: Answers and their percentages for questions to collect the user experience coming from students when interacting with MeIPeAS.

Question	Answer 1	Answer 2
1	Yes (40%)	No (60%)
2	Yes (80%)	No (20%)
3	Yes (25%)	No (75%)
4	Funny (94%)	Boring (6%)

Table 2: Answers and their percentages for questions to collect the user experience coming from teachers when seeing students interacting with MeIPeAS.

Question	Answer 1	Answer 2
1	Yes (100%)	No (0%)
2	Yes (80%)	No (20%)
3	Yes (80%)	No (20%)
4	Funny (100%)	Boring (0%)

5. Conclusions and Future Work

Today, the fundamental goal and purpose of having tutors (virtual support teachers) is a reality as an additional help to students in learning and obtaining knowledge. However, despite all the significant developments in a credible virtual tutor similar to a human teacher, there is still a large path of research to do. There are many interesting and sophisticated works about virtual tutors in support of education, either in the teaching of a new language (Forsyth et al., 2019), as a help in learning for children with disorders, such as autism (Tartaro and Cassell, 2008; Milne et al., 2018), among others, such as technologies for kindergarten children and tools to teach them to read (Cole et al., 2007).

By using agent technologies, it is possible to satisfy all the expected pedagogical requirements of virtual tutors. Although much research has been done in this domain, virtual tutors still need to be improved to meet all the expected requirements. These efforts can lead to the creation of tutors so credible and similar in behavior and appearance to that of a human teacher, and so sensitive and effective with their students to induce a better and greater motivation to learn and apply the knowledge obtained in the classroom. Experimental results suggest that virtual pedagogical agents that produce natural head movements and appropriate facial expressions while narrating a story produce much more positive user experiences than a virtual pedagogical agent that lack these behaviors (Yan and Agada, 2010).

Dincer and Doganay (2015) summarized that a pedagogical agent is generally described as an educational program that guides, motivates learners while encouraging them during learning by providing feedback. Studies in the field have revealed that such computer-assisted instruction programs in the form of agents increase the motivation of learners. In order to keep motivation levels high, these programs need to be adopted as per the individual needs. The, it can be beneficial to integrate the software design that can be personalized to the users' needs and application context and domain.

Mohanty (2016) listed the strengths and weaknesses of pedagogical agents as virtual tutors. The most relevant are as follows:

- Affective pedagogical agents have the ability to tutor a massive number of people, primarily constrained only by technology resources. These can boost feelings of self-efficacy.
- Promote individualized instruction, adaptive to user needs and facilitate people to treat computers like human tutors.
- The fantasy element of interacting with another tutor/agent enhances student motivation through use of body gestures, facial expressions, and voice intonation.

- Intelligent affective pedagogical agents with multiple functions are very complex to create.
- Natural language-understanding technology is still in its infancy and text-to-speech uses a robotic voice, which can often be annoying to learners.
- Speech recognition technology is not strong enough for widespread use.
- People treating the computer as their companion could become over-dependent and vulnerable to their negative effects.

Virtual environments for children's education both in Mexico and in other countries have been created since the last three decades. However, it is always interesting to watch how such works are addressed, so that the application areas can represent innovation and contribution to the research field (Sottilare et al., 2016). In this chapter, a novel and interactive prototype of a virtual tutor has been presented for use as a support tool in the teaching-learning process in the first three grades of primary education in Mexico. For this, activities were created in which students came to reinforce the knowledge that is taught within the classroom. In this way, it is possible to bring information and communication technologies closer to children in the form of a human-computer interaction solution.

The presented virtual tutor has to support the reinforcement of subjects taught in class, through exercises, which allow generation of social and emotional interaction, fostering a natural and real-time relationship with the student. To the authors' knowledge, MeIPeAS as a virtual tutor is unique and original in the pedagogical agent research field and one of the first created in Mexico for children's educational purposes. It is worth mentioning that in the first prototype attempts, the virtual tutor here presented was made known in the 2014 Mexican Space of Science and Technology: The Era of Information and Communication Technologies, during the workshop of Sociable Robots and Virtual Tutors on October 12, 2014. This event was held in the city of Toluca, the state of Mexico in Mexico. From that day, the prototype was perfected and extended year after year until achieving what was described and presented here. However, the research work is still in progress to concretise a final version to be released as soon as possible.

Proposing a prototype of an educational virtual tutor as a support tool, that contributes and positively impacts educational didactics at the primary level in Mexico, opens the way to reflect and imagine new scenarios of information and communication technologies application (Mellati and Khademi, 2019) in conjunction with the exploitation of serious games and autonomous agents within the educational field (Martha and Santoso, 2019; Payr, 2003). As work in the future, an extension of this prototype can be made to cover the other three missing grades of primary education in Mexico. In addition, other variants are possible to venture into for other levels of education as at the university level, where the subjects of study are more complex but are not exempt from being reinforced with the help of a virtual tutor. As in similar works, a future extension for the prototype exposed here includes use of the virtual tutor to provide support for self-explanation. This extension would be in terms of dialogues with students, where the agent asks questions to guide students in the teaching-learning process. Additionally, another extension is concerned with the use of a mixed short-term and long-term memory mechanism to create dialogues according to the progress of the students and what reminds the virtual tutor to provide customized feedback to them.

6. Future Trends

According to Pivec et al. (2004), every artefact, especially computers and even more so agents, are subject to anthropomorphization by the user. This means the user, to a certain extent, treats the machine like a human being. This is an argument for providing features to virtual teaching agents that are similar to the capabilities and features of human teachers. The guiding role, that can be found even in the most participative teaching styles, should be preserved for virtual teachers. In the same way, today's human-computer interaction still lacks the emotional component, giving the learner the feeling of being treated in an impersonal way.

Virtual characters have become ubiquitous in multimedia applications, appearing in numerous contexts, such as characters in video games, *avatars* in immersive worlds, or tutors in multimedia learning environments. One type of virtual character is known as a pedagogical agent, which is a life-like character present in multimedia learning environment in order to facilitate learning. Pedagogical agents have been researched for nearly three decades, yet the effectiveness of including an agent in a learning environment remains debatable. Schroeder and Gotch (2015) conducted a study to critically examine the persisting issues in pedagogical agent research that examines the use of an agent in a non-agent condition. The analysis of the literature highlights the main persisting issues in agent research, specifying an agent's instructional role, defining the agent's personal attributes and providing promising future research directions in science, technology, engineering, and mathematics (STEM) in classrooms as well.

Mohanty (2016) mentions that very little research has been done to address the details of how human tutors adapt to particular affective signals in particular teaching environments. Hence research has to be directed towards designing an artificial affective tutor/pedagogical agent, who would respond to the students' emotions and accordingly guide and motivate them towards better learning. In addition, it is argued that in real-life physical learning-environment, students' feedback is a major requirement that the human tutors use to constantly adapt their teaching strategies. Moreover, human tutors use the affective information they receive from learners to individualize the tutoring sessions so that learning is as effective and efficient as possible for each student/learner. The affective feedback from students that human tutors adapt can take many forms, such as facial expressions, gestures and vocal inflections, mostly non-verbal or para-linguistic (affective communication) in nature. Due to this non-verbal, voice modulation and spontaneous/involuntary responses, the human tutors constantly receive a significant amount of affective feedback during tutoring sessions, which convey the moods and emotions of the learner, whether he/she is happy, confused, frustrated, surprised, bored or content with the learning materials being presented to him/her. Mostly human tutors use this information to adapt their teaching to match the affective state of the learner. Thus, it implies that capable intelligent tutors, who could adapt to students' emotions, could also significantly enhance their learning and performance.

The possible multi-disciplinary research to invent pedagogical agents that behave like sensitive and effective human teachers should address the main technological challenges and research breakthroughs required to create human-like virtual tutors. It is well known that students generally expect virtual tutors to be believable as human teachers or mentors, be entertaining, easy to communicate, helpful and diversified. At least, there exist five issues in which research is needed to satisfy learner desires for pedagogical agents: believability, emotionality, team working, personalization and entertainment. Research on the effectiveness of advanced learning technologies (ALTs) on ubiquitous learning suggests that students are often poor at self-regulating their learning and, as a consequence, researchers have implemented pedagogical agents to help foster students' use of cognitive, affective, metacognitive and motivational self-regulated learning (SRL) processes. Several analyses have been conducted that investigated the impact and effectiveness of pedagogical agents on learning with ALTs. These results indicated that the effectiveness of pedagogical agents in ALTs is dependent on many factors, such as the content under study, the population of students, and the features of the pedagogical agents themselves. Thus, it is important to consider all these details when designing and measuring the effectiveness of pedagogical agents in ALTs (Taub et al., 2016).

References

Abbot, C. (2003). ICT: Changing Education, Routledge.

Abdullah, A., Adil, M., Rosenbaum, L., Clemmons, M., Shah, M., Abrahamson, D. and Neff, M. (2017). Pedagogical agents to support embodied, discovery-based learning. 2017 International Conference on Intelligent Virtual Agents, pp. 1–14, Stockholm, Sweden: Springer, Cham.

AlKhayat, A. (2017). Exploring the effectiveness of using chatbots in the EFL classroom. pp. 20–36. *In*: Hubbard, P. and Ioannou-Georgiou, S. (eds.). Teaching English Reflectively with Technology. Kent, UK: IATEFL.

Almasri, A., Ahmed, A., Almasri, N., Abu Sultan, Y.S., Mahmoud, A.Y., Zaqout, I.S. and Abu-Naser, S.S. (2019). Intelligent tutoring systems survey for the period 2000–2018. International Journal of Academic Engineering Research, 3(5): 21–37.

Ashoori, M., Miao, C. and Goh, E.S. (2009). Toward a Model of Intelligence in Pedagogical Agents.

Atkinson, R.K., Mayer, R.E. and Merrill, M.M. (2005). Fostering social agency in multimedia learning: Examining the impact of an animated agent's voice. Contemporary Educational Psychology, 30(1): 117–139.

Azevedo, R., Martin, S.A., Taub, M., Mudrick, N.V., Millar, G.C. and Grafsgaard, J.F. (2016). Are pedagogical agents' external regulation effective in fostering learning with intelligent tutoring systems? 2016 International Conference on Intelligent Tutoring Systems, pp. 197–207, Zagreb, Croatia: Springer, Cham.

Bächle, M., Daurer, S., Judt, A. and Mettler, T. (2018). Chatbots as a User Interface for Assistive Technology, uDay XVI, pp. 1–10, Laussane, Switzerland: University of Laussane.

Baker, R.S. (2016). Stupid tutoring systems, intelligent humans. International Journal of Artificial Intelligence in Education, 26(2): 600–614.

Bascopé, M., Perasso, P. and Reiss, K. (2019). Systematic review of education for sustainable development at an early stage: Cornerstones and pedagogical approaches for teacher professional development. Sustainability, 11(13): 719.

Bhattacharjee, B. and Deb, K. (2016). Role of ICT in 21st century's teacher education. International Journal of Education and Information Studies, 6(1): 1–6.

Blair, K., Schwartz, D.L., Biswas, G. and Leelawong, K. (2007). Pedagogical agents for learning by teaching: Teachable agents. Educational Technology, 47(1): 56–61.

Boudourides, M. (2003). Constructivism, education, science, and technology. Canadian Journal of Learning and Technology/La revue canadienne de l'apprentissage et de la technologie, 29(3).

Carperter, R. (2019). Cleaverbot, Retrieved 09 07, 2019, from https://www.cleverbot.com/.

Chatbots.org. (n.d.). Retrieved 09 06, 2019, from http://chatbots.org.

Cole, R., Wise, B. and van Vuuren, S. (2007). How Marni teaches children to read. Educational Technology, 47(1): 14–18.

Cuéllar, G.R. (2012). La Reforma Integral de la Educación Básica en México (RIEB) en la educación primaria: Desafíospara la formacióndocente. Revistaelectrónica inter universitaria de formación del profesorado, 15(1): 51–60.

Dincer, S. and Doganay, A. (2015). The impact of pedagogical agent on learners' motivation and academic success. Practice and Theory in Systems of Education, 10(4): 329–348.

Existor Ltd. (2016). Eviebot, Retrieved 09 07, 2019, from https://www.eviebot.com/en/.

Existor Ltd. (2018). Pewdiebot. Retrieved 09 07, 2019, from https://www.pewdiebot.com/en/.

Forsyth, C.M., Luce, C., Zapata-Rivera, D., Jackson, G.T., Evanini, K. and So, Y. (2019). Evaluating English language learners' conversations: Man vs. machine. Computer-assisted Language Learning, 34(2): 398–417.

Guichard, S. (2005). The Education Challenge in Mexico: Delivering Good Quality Education to All, OECD Economics Department Working Papers, No. 447, Paris, France: OECD Publishing.

Hautala, J., Baker, D.L., Keurulainen, A., Ronimus, M., Richardson, U. and Cole, R. (2018). Early science learning with a virtual tutor through multimedia explanations and feedback on spoken questions. Educational Technology Research and Development, 66(2): 403–428.

Huang, R., Spector, J. and Yang, J. (2019). Learner experiences with educational technology. pp. 91–105. *In*: Huang, R., Spector, J. and Yang, J. (eds.). Educational Technology: Lecture Notes in Educational Technology. Singapore: Springer.

Imbert, R., Sánchez, L., De Antonio, A., Méndez, G. and Ramírez, J. (2007). A multi-agent extension for virtual reality based intelligent tutoring systems. Seventh IEEE International Conference on Advanced Learning Technologies (ICALT 2007), pp. 82–84, Niigata, Japan: IEEE.

Johnson, W.L., Rickel, J.W. and Lester, J.C. (2000). Animated pedagogical agents: Face-to-face interaction in interactive learning environments. International Journal of Artificial Intelligence in Education, 11(1): 47–78.

Johnson, W.L. and Lester, J.C. (2016). Face-to-face interaction with pedagogical agents, twenty years later. International Journal of Artificial Intelligence in Education, 26(1): 25–36.

Johnson, W.L. and Lester, J.C. (2018). Pedagogical agents: back to the future. AI Magazine, 39(2): 33–44.

Khanna, A., Pandey, B., Vashishta, K., Kalia, K., Pradeepkumar, B. and Das, T. (2015). A study of today's A.I. through chatbots and rediscovery of machine intelligence. International Journal of u- and e-Service, Science and Technology, 8(7): 277–284.

Kim, Y. and Baylor, A.L. (2016). Research-based design of pedagogical agent roles: A review, progress, and recommendations. International Journal of Artificial Intelligence in Education, 26(1): 160–169.

Kulik, J.A. and Fletcher, J.D. (2016). Effectiveness of intelligent tutoring systems: A meta-analytic review. Review of Educational Research, 86(1): 42–78.

Lester, J., Branting, K. and Mott, B. (2004). Conversational agents. *In*: The Practical Handbook of Internet Computing, Chapman & Hall/CRC.

Loebner Prize. (n.d.). Retrieved 09 06, 2019, from http://www.loebner.net/Prizef/loebner-prize.html.

Martha, A.S. and Santoso, H.B. (2019). The design and impact of the pedagogical agent: a systematic literature review. Journal of Educators Online, 16(1): 1–15.

Masche, J. and Le, N.T. (2017). A review of technologies for conversational systems. 2017 International Conference on Computer Science, Applied Mathematics and Applications, pp. 212–225, Berlin, Germany: Springer, Cham.

Mayer, R.E. (2005). Cognitive Theory of Multimedia Learning. The Cambridge Handbook of Multimedia Learning. New York, USA: Cambridge University Press. 41: 31–48.

McLaren, P. (2015). Life in Schools: An Introduction to Critical Pedagogy in the Foundations of Education, London UK, New York USA: Routledge.

Mellati, M. and Khademi, M. (2019). Technology-based education: Challenges of blended educational technology. pp. 48–62. *In*: Habib, M. (ed.). Advanced Online Education and Training Technologies, IGI Global.

Michael, D.R. and Chen, S.L. (2005). Serious Games: Games that Educate, Train, and Inform, Muska & Lipman/Premier-Trade.

Milne, M., Raghavendra, P., Leibbrandt, R. and Powers, D.M. (2018). Personalization and automation in a virtual conversation skills tutor for children with autism. Journal on Multimodal User Interfaces, 12(3): 257–269.

Mitrovic, A. and Suraweera, P. (2000). Evaluating an animated pedagogical agent. International Conference on Intelligent Tutoring Systems, pp. 73–82, Montreal, Canada: Springer, Berlin, Heidelberg.

Mohammadhasani, N., Fardanesh, H., Hatami, J., Mozayani, N. and Fabio, R.A. (2018). The pedagogical agent enhances mathematics learning in ADHD students. Education and Information Technologies, 23(6): 2299–2308.

Mohanty, A. (2016). Affective pedagogical agent in e-learning environment: A reflective analysis. Creative Education, 7(4): 586–595.

Moreno, J.C. and Toro, C.H. (2010). BotBlog: A blogs and agents integration proposal. Sistemas y Telemática, 8(15): 51–65.

Morton, H. and Jack, M.A. (2005). Scenario-based spoken interaction with virtual agents. Computer-assisted Language Learning, 18(3): 171–191.

Pandorarobots. (n.d.). Meet Mitsuku. Retrieved 09 06, 2019, from https://www.pandorabots.com/mitsuku/.

Passey, D., Dagienė, V., Atieno, L.V. and Baumann, W. (2018). Computational practices, educational theories and learning development. Problemos, 24–38.

Payr, S. (2003). The virtual university's faculty: An overview of educational agents. Applied Artificial Intelligence, 17(1): 1–9.

Pelgrum, W.J. (2001). Obstacles to the integration of ICT in education: results from a worldwide educational assessment. Computers & Education, 37(2): 163–178.

Pivec, M., Baumann, K. and Gütl, C. (2004). Everything virtual—Virtual classes, virtual tutors, virtual students, virtual emotions—but the knowledge. World Conference on Educational Multimedia, Hypermedia and Telecommunications, pp. 4009–4015, Lugano, Switzerland: Chesapeake.

Schroeder, N.L. and Adesope, O.O. (2015). Impacts of pedagogical agent gender in an accessible learning environment. Journal of Educational Technology & Society, 18(4): 401–411.

Schroeder, N.L. and Gotch, C.M. (2015). Persisting issues in pedagogical agent research. Journal of Educational Computing Research, 53(2): 183–204.

Scott, D., Posner, C., Martin, C. and Guzman, E. (2018). The Education System in Mexico, UCL Press.

Song, D. (2017). Designing a teachable agent system for mathematics learning. Contemporary Educational Technology, 8(2): 176–190.

Sottilare, R., Graesser, A., Hu, X. and Brawner, K.R. (2016). Design Recommendations for Intelligent Tutoring Systems: Authoring Tools and Expert Modeling Techniques, Orlando, Florida, USA: US Army Research Laboratory.

Taoum, J., Nakhal, B., Bevacqua, E. and Querrec, R. (2016). A design proposition for interactive virtual tutors in an informed environment. International Conference on Intelligent Virtual Agents, pp. 341–350, Los Angeles, USA: Springer, Cham.

Tartaro, A. and Cassell, J. (2008). Playing with virtual peers: bootstrapping contingent discourse in children with autism. 8th International Conference on International Conference for the Learning Sciences, 2: 382–389, Utrecht, The Netherlands: ACM.

Taub, M., Martin, S.A., Azevedo, R. and Mudrick, N.V. (2016). The role of pedagogical agents on learning: Issues and trends. pp. 362–386. *In*: Mendez Neto, F.M., de Souza, R. and Gomes, A.S. (eds.). Handbook of Research on 3-D Virtual Environments and Hypermedia for Ubiquitous Learning, Hershey, Pennsylvania, USA: IGI Global.

Treviño Ronzón, E. and Cruz Vadillo, R. (2014). La Reforma Integral de la EducaciónBásica en el discursodocente: Análisisdesde el ángulo de la significación. Perfileseducativos, 36(144): 50–68.

Van Manen, M. (2016). The Tact of Teaching: The Meaning of Pedagogical Thoughtfulness, London UK, New York USA: Routledge.

Wallace, R. (2009). The anatomy of A.L.I.C.E. pp. 181–210. *In*: Epstein, R., Roberts, G. and Beber, G. (eds.). Parsing the Turing Test, Springer, Dordrecht.

Ward, W., Cole, R., Alonso, D.B., Martin, C.B., Svirsky, E. and Weston, T. (2013). My science tutor: A conversational multimedia virtual tutor. Journal of Educational Psychology, 105(4): 1115–1125.

Wise, B., Cole, R., van Vuuren, S., Schwartz, S., Snyder, L. and Ngampatipatpong, N. (2005). Learning to Read with a Virtual Tutor: Foundations Literacy, Interactive Literacy Education: Facilitating Literacy Environments Through Technology, Mahwah, New Jersey, USA: Lawrence Erlbaum.

Worswick, S. (n.d.). Mitsuku Chatbot. Retrieved 09 06, 2019, from http://www.square-bear.co.uk/mitsuku/home.htm.

Wouters, P., Van Nimwegen, C., Van Oostendorp, H. and Van der Spek, E.D. (2013). A meta-analysis of the cognitive and motivational effects of serious games. Journal of Educational Psychology, 105(2): 249–265.

Yan, J. and Agada, R. (2010). Life-like animated virtual pedagogical agent enhanced learning. Journal of Next Generation Information Technology, 1(2): 4–12.

Chapter **7**

Gamification in Virtual Reality Environments for the Integration of Highly Effective Teams

Mirna Muñoz,[1] *Adriana Peña Pérez Negrón*[2,*] and *Luis Hernández*[1]

1. Introduction

Nowadays, software development is carried out by teams whose performance is affected by several factors. A particularly important factor in teamwork is to achieve the integration of the persons that compose it, since it is strongly linked to the level of team cohesion and performance. It is also essential that all professionals have the ability to work as a team, especially in the area of software engineering.

Based on the above, the proper integration of a work team influences its performance; the skills, knowledge and interactive styles of each member must be complementary, resulting in a highly effective team. This situation highlights the importance of focusing on the human factor to help organizations reduce the risk of failure on account of the teams' integration problems.

The activities to integrate a team are important because they can strengthen both personal and professional relationships among its members. Today, there are several activities to help carry out the integration of a team, with the aim of improving communication, coordination and teamwork. They are a great support in improving their performance independently of the area.

In this context, this research presents a virtual environment containing the gamification elements as a tool to identify the roles in a software development team through the construction of a virtual reality environment based on the Rube Goldberg Machine, the support of gamification techniques, and the activities described in the Time-

---

[1] Centro de Investigación en Matemáticas, Mexico.
[2] Universidad de Guadalajara, Mexico.
* Corresponding author: adriana.pena@cucei.udg.mx

Space-Position information (TSPi) methodology which presents an attractive way to integrate software development teams.

Software development is not only programming a system, but is a complex task that nowadays is performed by teams in both industrial and academic domains. In this context, there are factors that impact the success or failure of software development projects, such as clarification of goals, methodology used, planning assurance, team integration and team performance, in which the human factor is most relevant because the abilities and knowledge of each member are directly related to the team success.

To achieve an optimal level of cohesion in a team, it is often necessary for the team to spend large periods of time together as this time is critical for the project's success and for achievement of the project's goals. Furthermore, features, such as team members' personality, knowledge, and abilities are important factors that make a high impact upon how well a team integrates and performs (Dorling and McCaffery, 2012).

For this reason, it is important to focus on solving the problems related to human factors, such as team communication or team work in order to carry out the software development projects, which at times is even more important than technical knowledge for team success (Muñoz et al., 2016).

Then, it is fundamental that the team members have the required skills that enable teamwork, such as communication, personal relationships skills, the capacity to solve problems in order to perform the activities assigned in a project in a proper way, so that the project goals can be fulfilled on time and in the expected way.

Nowadays, the most common way to assign a team role is through surveys or focusing on the people's experience, but that most of the time, it results in an inadequate role assignment. Moreover, there is a lack of software tools for integrating software development teams furthermore for integrating teams.

To provide a solution, this research focuses on providing a new and adequate way to integrate teams in the software development field, using four technologies: virtual reality environments, gamification, interactive styles, and team software-process methodology.

- Virtual reality environments are computer-generated 3D spaces that compel the users to get the feeling of being in an environment other than the one they actually are in, and to interact with that environment (Ellis, 1995).

- Gamification in the field of software engineering has been applied in software development to help in the achievement of activities as well as to motivate and improve communication skills (Knutas et al., 2014).

- Interactive styles refer to a metric of the natural preferences of each person related to the way in which an individual behaves in specific situations (Muñoz et al., 2017).

- Team software process is a methodology for software development that provides a set of software development phases and a set of team roles, including scripts, and also a set of activities to be carried out in each phase and by each role (Muñoz et al., 2017).

After the introduction, the rest of the paper is composed of Section 2, which shows the background to this research, that includes key concepts involved in this research as well as related works; Section 3 shows the selection of technology regarding the virtual reality environment requirements; Section 4 describes the steps followed to develop the software tool; Section 5 presents the software tool; and finally, Section 6 provides the discussion, conclusions, and future work.

2. Research Background

2.1 Key Concepts

2.1.1 Virtual Reality Environments

Virtual reality (VR) is a powerful context in which time, scales, and physics can be controlled, where participants can acquire entirely new capabilities, such as flying, having an object as a virtual body, or observing the environment from different perspectives. In virtual environments materials do not break or wear out, allowing safe experiences in, for example, distant or dangerous locations (Bricken, 1991).

On its multiuser approach, Collaborative Virtual Environments (CVE), VR compels the users to feeling the co-presence, allowing them to interact with each other (Schroeder, 2006) through a graphic representation, their *avatar*. For spatial tasks, the CVE's predominantly visual characteristic is valuable, especially for object manipulation and non-verbal communication (Spante et al., 2003).

Computers have always been linked to learning and training. Beyond, its well-known motivational impact, in VR, apprentices may feel less pressured than in real-life scenarios as they get the opportunity of understanding consequences of their choices without suffering those (Peña and Jiménez, 2012).

This generated state of mind on the users makes VR a promising environment for the study of people's behavior. CVE allow researchers to create experimental situations, more realistic than controlled laboratory setting, without interference and with the possibility of automatic collection of data without the user feeling observed. Furthermore, it is a proper media in which gamification elements can be easily applied.

2.1.2 Gamification

The term gamification refers to the use of game elements in a non-gaming environment (Steffens et al., 2015). The main objectives of gamification are to motivate people to achieve their personal goals, to change their behavior, to develop new skills (Burke, 2014). This is the reason to present gamification as a new strategy to influence and motivate people during the development of activities that they are not used to perform or those in which they try to acquire a new habit.

Gamification is not a new term; it emerged in the late 2010 and early 2011. Its first applications took place on varied domains such as: (1) health, where applications to perform physical activities were created by using gamification, and (2) social interaction and learning, where applications containing mainly points and position tables were created.

The use of gamification has increased since 2010, so that looking for the gamification concept in Google trends, in a period from 2010 until October 2018, as Fig. 1 shows, presents a trend of constant increase in the subject. The data in Fig. 1 was extracted from: https://trends.google.com.mx/trends/explore?date=2010-01-12%20 2018-09-30&q=gamification.

The areas in which gamification have mainly been applied are marketing, finance, human resources, environment, government, and mobile applications with different objectives. One of the most famous examples of the application of gamification is the speed camera lottery of the fun theory—an initiative carried out by Volkswagen by the

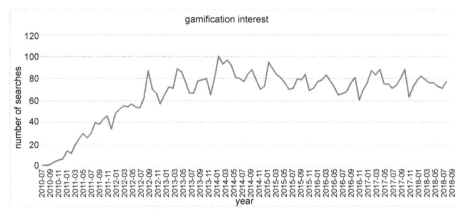

Fig. 1: Trend in gamification search.

end of 2010 (Volkswagen, 2010). This proposal aimed to make drivers aware of reducing speed in one of the main avenues in the city of Stockholm. It took photos of the drivers who respected the speed limit signs; if the driver respected the speed, a system would take a picture of the car to register it in a raffle. Then, the drivers who emerged winners received financial compensation for their proper behavior. As a result, the drivers respected the speed limit signs (Volkswagen, 2010).

Besides, within frameworks addressing gamification topics, there are four authors that proposed the use of gamification elements from different perspectives and with different compositions:

The framework proposals of Burke (2014) and of Werbach and Hunter (2012) are focused on the business field. They proposed to apply six different steps to implement gamification in organizations. On the one way, the composition of Burke's framework is: (1) results of business and success metrics; (2) target audience; (3) player objectives; (4) commitment model; (5) game and travel space; and (6) game economy. He also mentioned play, prove, and iterate. On the other way, the composition of Werbach and Hunter's framework is: (1) define business goals; (2) describe players; (3) define behavior; (4) design cycles and activities; (5) do not forget the fun; and (6) implement the appropriate tools.

The framework proposed by Kapp (2012) is focused on providing a reference framework to understand the way of applying gamification techniques. Its goal is the creation of activities that make knowledge available to the participants. This framework is composed of a set of instructions, steps, and recommendations to carry out the proposal of gamification activities. The framework is applied to the educational domain.

Finally, the framework proposal of Chou (2015) focuses on applying the gamification elements based on the final users. In this way, the application of gamification elements allows coupling to the end-user environment regardless of the area of application. This framework is composed of eight areas that the author names as 'kernels'. Each area represents a particular objective to be addressed by the gamification proposal: (1) epic sense; (2) development and commitment; (3) empowerment of creativity and feedback; (4) ownership and possession; (5) social influence and relationship; (6) shortage and impatience; (7) unpredictability and curiosity; and (8) loss and evasion.

2.1.3 *Interactive Styles*

An interactive style is the way in which an individual faces a particular situation or problem. It is analyzed in such a way that allows evaluating the differential behavior of each individual (RibesIñesta, 2009). This research work is based on the concept of interactive styles and types of interactive styles proposed by RibesIñesta (1990).

An interactive style is defined on the basis of three criteria:

- How an individual has reacted in past situations, which are part of his/her experience.
- Description of the situation to face.
- Prediction of how an individual will react to future situations.

The process to identify an interactive style is composed of four steps:

1. *Responses in historical situations*: The identification of an interactive style can be based on the responses that an individual has presented in previous situations, where he/she has applied knowledge, skills, or experience to solve a situation or a particular problem.
2. *Description of a characteristic interactive mode*: The identification of an interactive style can be made from the description of a situation that the individual must face.
3. *Identification of the interactive style*: To continue with the identification of an interactive style, the individual must be exposed to the situation or problem described in Step 2. At this point, enough information must be collected to perform the evaluation of the individual with respect to the interactive style. Once the individual has finished his/her participation in the situation, the identification of the interactive style can be performed.
4. *Conduct of an individual in a future situation*: The obtained results, based on the individual answers, can be used to predict the behavior or a series of responses of the individual in future situations or problems.

Within the definition of interactive styles, there are two concepts that must be defined to understand the functionality and way of application. Both are taken as the base in the evaluation of an interactive style:

- *Close contingency*: It is based on giving a person a set of indications to carry out an activity; in this way, it is possible to evaluate if and how the person follows the indications.
- *Open contingency*: It is based on the absence of indications to carry out an activity, so that the person must resolve the activity without indications regarding the goal to achieve. In this way it is possible to evaluate how the person carries out the activity based on his/her experience or previous knowledge.

This research work addressed 12 interactive styles. These interactive styles have been studied by RibesIñesta (1990) and previously interpreted and described in Rangel et al. (2017b) and Muñoz et al. (2018).

1. *Making decisions*: It refers to the possibility of getting only one response to stimulus that are competitive or uncertain in time; in other words, how an individual selects a logical choice from the available situations.

2. *Ambiguity tolerance*: It refers to the responses under functional antagonist proprieties, and/or differences between signals and supplementation; in other words, how an individual responds in a regular affected way to doubtful or uncertain situations.

3. *Frustration tolerance*: It refers to performing under conditions with no signals of interference, diminishing, loss, or delay in consequences; in other words, how an individual keeps his/her performance under not clear signals of the consequence.

4. *Achievement or persistency*: It refers to performing under conditions of growing or higher requirements; in other words, how an individual keeps his/her performance under conditions of higher or growing requirements.

5. *Flexibility to the change*: It refers to change in response to facing signaled, not signaled, or unspecific contingencies; in other words, how an individual is able to adapt his/her responses to face different contingencies.

6. *Tendency to transgression*: It refers to give a response to a not-response signal; in other words, how an individual gives a response without following rules.

7. *Curiosity*: It refers to giving diverse responses to contingencies that do not require them; in other words, how an individual provides a response in the case in which it is not required.

8. *Tendency to risk*: It refers to choose contingencies signaled as those with a mayor possibility of gain or loss in contrast with constant alternatives; in other words, how an individual prefers to select a situation with most possibility to win or lose in contrast to those situations with constant alternatives.

9. *Dependence to signs*: It refers to adjust responses that have redundant signs in a contingency; in other words, how an individual is able to adjust his/her responses to redundant signs in the contingency.

10. *Responsively to new contingencies and signs*: It refers to how a response can be affected when facing new signs of the same contingency or in the case of a new contingency with the same sign; in other words, how an individual affects his/her response when facing new signs in the same situation or a new situation with the same signs.

11. *Impulsivity*: It refers to corresponding responses in those situations where the contingencies have the feature where their components are not homogenous; in other words, how an individual gives a response in an emotional or spontaneous way.

12. *Conflict reduction*: It refers to a response to the concurrent opposites or to competitive signs; in other words, how an individual gives a response facing to two types of signs: concurrent opposites or competitive.

2.1.4 Team Software Process Introduction

To get a guideline of activities to be performed by a team to develop software, this research work selected the TSPi methodology (Humphrey, 2006). TSPi is an academic version of the TSP, which guides engineers in the creation and maintenance of self-managed software-development teams, allowing engineers to set goals and commitments so that they are in charge to plan and monitor their work (Hernández-López et al., 2010).

TSPi provides in a clear and accurate way the different phases to be performed by a team in the software development process. Besides, it provides a definition of the main roles of the participants throughout the software development process, providing a description of their activities (Humphrey, 2006). Next, the phases and roles proposed by TSPi are briefly described.

TSPi provides a software development process that contains eight phases (Humphrey, 2006):

- *Launch*: The purpose of this phase is to constitute the work team, assign the roles and explain the objectives that must be achieved in the product development.

- *Strategy*: The purpose of this phase is to create a development strategy to build the product, estimate the project hours, produce the conceptual design and the risk assessment.

- *Plan*: The purpose of this phase is to produce the project's activity schedules, both individual and as a team. This calendar must be made to the level of activities and the time to fulfill them. Besides, the quality plan should be developed, including the quality criteria and objectives to be met by the team.

- *Requirements*: The purpose of this phase is to produce the Software Requirements Specification (SRS), where the product functionality and the case-study description for a normal or abnormal operation of the product are defined.

- *Design*: The purpose of this phase is to produce the Software Design Specification (SDS), where the product components and their specifications are described. In addition, the conventions for the nomenclature of the files are defined and an integration testing plan is devised.

- *Implementation*: The purpose of this phase is to carry out the product code. Once codified, the compilation of each component is made and at the same time the established quality objectives for the product are verified.

- *Testing*: The purpose of this phase is to inspect the end product by carrying out the monitoring of testing plans for the unit test, the components test and the system test, in order to evaluate the quality of the developed product. In addition, the user documentation is made.

- *Postmortem*: The purpose of this phase is to evaluate the obtained results throughout all software development phases while identifying the team's performance and redacting the report of the project's life cycle.

TSPi focuses on six roles (Humphrey, 2006):

- *Team leader*: He/she is the project manager, in charge of building, maintaining, and motivating effective teams. He/she should analyze the capabilities and abilities of team members to achieve a successful project. In addition, he/she is in charge of presenting to the management the actual state of the project.

- *Quality/process manager*: He/she is in charge of ensuring that the TSPi process is followed, as well as for verifying the quality of teams' products. He/she leads the team during the development of a thorough plan to be used to analyze the quality of the project and prevent any potential issues.

- *Planning manager*: He/she is in charge of producing a complete plan for the team, and managing and monitoring it. He/she should help the team members to develop their personal plans and to monitor the progress of the assigned tasks.

- *Development manager*: He/she is in charge of creating the product itself, based on the software architecture. He/she guides the team in the definition, design, and development of the product tests.
- *Support manager*: He/she is in charge of carrying out the configuration management and risk management, so that he/she helps the team to establish, get, and manage the technological tools and methods needed to perform the project.
- *Engineer*: All members should participate in the development of the software product.

2.2 Related Works

This section provides an overview of related works in which virtual environments together with gamification for teams have been used. It is important to mention that these related works were selected from a systematic review published by (Hernández et al., 2017), highlighting only those related works that implement a virtual environment, a serious game or platforms for teamwork.

Ellis et al. (2008) created a set of games for construction activities in distributed virtual teams. They mention that having a great number of communications is not related with good communication. It could indicate an absence of clarity in the idea interchange. To construct the games, they use three types of activities: a puzzle, a tower construction, and a castle design and construction. They conclude that a not-proper role assigned to teams entails a negative effect in both work coordination and team development.

Herranz et al. (2015) presented how, through a web platform, the gamification concept can be applied to the development of software. The use of the platform improves the employees' participation in the software improvement processes. They aim to study the impact of this tool on motivation. They present a case study whose results show an increment in the employees' participation in the software improvement processes.

Häkkinen et al. (2012) presented a proposal for eSpace game to solve puzzles. This proposal aims to integrate the team gradually, improving its communication skills. The results of this proposal show that the game can be useful to build teams and evaluate leadership in enterprises and also to create team leaders.

Wendel et al. (2013) presented the design of a serious game, which aims to improve the soft skills in a team, such as communication, collaboration, motivation, and teamwork. The serious game focuses on solving a set of puzzles, each with a higher degree of difficulty. The results show that participants collaborate during the game. Besides, the use of digital multiuser-videogames motivates the trust and cooperative behavior among team members.

Lukosch et al. (2014) presented a proposal for a virtual environment designed to improve soft skills of teamwork. This proposal aims to explore the use of serious games for supporting activities, such as communication, effective cooperation, decision-making, interpretation, and processing of information. The results of this proposal show that the virtual environment improves the teamwork's abilities, especially communication skills.

Guenaga et al. (2014) presented a proposal of a serious game for building distributed virtual teams. This proposal aims to improve communication among team members in order to achieve effective communication. The results of this proposal show that the roles assigned impact the team coordination and performance. Besides, it allows knowing the impact of gamification in teamwork.

Bozanta et al. (2016) presented a proposal for the use of serious games as an effective tool to improve the team members' cohesion, commitment, communication, and collaboration. This proposal aims to explore the effect of serious games on the cohesion of a team in multi-user virtual environments. The results show that the interface impacts the communication, collaboration, and performance of the team in virtual environments.

Benefield et al. (2016) created a number of models to predict team effectiveness. Their models include variables to control the team, intergroup closures, intergroup remunerations, and leadership. According to them, a team's performance is based on three factors: (1) standards to get quality instead of quantity; (2) social norms to support team members' interdependence; and (3) create learning positive experiences.

From the eight related works, only two related works focus on team integration; the rest of them focus on improving teamwork activities, once the team is built. This finding highlights the contribution of this research work because it focuses on team building.

3. Technology Selection

This section describes an analysis of development technologies to carry out the construction of the virtual environment for the identification of work profiles for a team, as a software tool. In order to select the technology, a comparison of the main graphic engines was made.

3.1 Comparison of Game Engines

A game engine can be defined as a software tool that supports the development of a video game (Polančec and Mekterović, 2017). Besides, Ali and Usman (2017) define a game engine as the set of software modules that are joined to form the basis for the development of computer games, based on virtual environments.

Based on the above mentioned, the main features that a game engine should cover are: (1) manipulation for physical properties of objects (weight, dimension, and collisions); (2) rendering; (3) sound; and (4) support of network development.

To make the comparison of game engines, a papers' search was carried out, focusing on 'game engines', 'analysis' and 'comparison' in the databases IEEE Xplore Digital Library and Springer Link. On the one hand, in the IEEE Xplore Digital Library the string 'Game engine analysis' received 84 results. On the other, in the Springerlink, the string 'Game Engine Analysis Comparison' for the computer science field obtained a result of 587.

The selection of the main studies related to the game engine topic was made applying the next criteria: (1) studies where the term 'game engine' was in the title or as part of the keywords; (2) studies in which a case study was performed; or (3) studies that performed an analysis for the game engine. After the application of the three criteria, a total of 13 main studies was selected.

Therefore, to extract the data related to the technology, the next steps were applied:

1. Read each selected study to identify its main literature contribution.
2. Identify the game engines analyzed in the study.
3. Identify the best game engine analyzed in the study.
4. Select the game engine for this research proposal.

As Table 1 shows, the best-rated engine for video games development was two: The Unity 3D and the Unreal Engine.

Unity 3D was rated as the best engine because it present the following features: (1) support for network development (Abdullah et al., 2014); (2) has multiple functionalities to offer, such as plug-in packages; application in the gaming market; compatibility with various development platforms, low cost; and less development time (Kim et al., 2014); (3) it could be implemented in multiple platforms (Christel et al., 2012); and (4) facilitates the construction of graphic elements on Windows, Mac and Linux systems. Besides, the use of OpenGL, OpenGL ES and WebGL on Unity 3D allows compatibility with Android, IOS and via web connection (Messaoudi et al., 2016).

Based on the obtained results and according to the virtual environment requirements, such as allowing modeling a set of different objects with specific features in weight,

Table 1: Shows the results of analysis performed by the game engines.

Study	Game Engines Analyzed	Best Rated Game Engine
Pavkov et al., 2017	Adventure Game Studio, Construct 2, e-Adventure, Game Maker: Studio, Phaser Editor	Game Maker: Studio
Ali and Usman, 2017	CE4, Id Tech5, Source, Unity 3D, Unreal Engine 3, Unreal Engine 4, Blender, Cocos 2D, Delta 3D, GMS, GS, Id Tech4, Quake 4, AFP, AT, Chrome, Construct 2, Dunia 2, Dx Studio, Ogre 3D, SIO2, Torque, Unigine, Frost Bite	Unity 3D
Söbke and Streicher, 2016	Unity, CryEngine, Game Maker, Source Engine, Unreal Engine, Glass Box Engine	None
Messaoudi et al., 2016	Unity 3D	Unity 3D
Westhoven and Alexander, 2015	CryEngine, Unity	CryEngine, Unity
Vasudevamurt and Uskov, 2015	Unity SIM, Torque, Unreal Engine, Ungine SIM, Neoazis, CryEngine, Game Maker Studio, Adobe Flash Professional, Neoaxis Engine, Game Salad, Cocos 2D, Quake, Construct 2, Shiva 3D, Delta 3D, Source Engine, Frost Bite, Snow Drop, Dunia 2, Fox, Chrome Engine 6	3D development: Unity, Unreal Engine 2D development: Adobe Flash Professional
Schweiger et al., 2014	Quintus, Crafty, Panda JS, Enchant JS, Lime JS, Jumru 5s	Jumru 5s
Uskov and Sekar, 2014	Unity Pro, Game Salad Creator, DX Studio, v-Play	Unity Pro, DX Studio
Abdullah et al., 2014	Unity 3D, ShiVa Engine, Irrlicht 3D Engine, Reality Factory, Panda 3D	Unity 3D
Kim et al., 2014	SIMDiS, VR-Vanatage, OSG, Ogre3D, Delta3D, Vega, Unity 3D	Unity 3D
Christel et al., 2012	Unity 3D	Unity 3D
Noh et al., 2006	Quake, Unreal Engine, Jupiter	Unreal Engine
Andreoli et al., 2005	Half-Life2, Unreal Engine 3, Torque Engine	Unreal Engine

size and form, so that they will be used within the construction of many chain reactions for achieving an objective and enabling use of gamification elements within the virtual environment.

The game engine Unity 3D was selected to develop the virtual framework. The main motivations for selecting this graphic engine are: (1) provides a videogame engine to manage the objects built in the environment; to manage the lights and shadows; to manage and simulate the physical properties of the objects; and to manage the objects animations; (2) supports third-dimension applications, so that, it will help the construction of the virtual environment; (3) performs the package of games in different platforms, such as PC (Linux, Mac, and Windows); mobile devices (Android, IOS); and Web platforms; and (4) support network development.

4. Methodology for Developing the Software Tool

4.1 Virtual Environment to be Developed

To develop the virtual environment, a set of four steps was performed as shown in Fig. 2.

1. *Identify the activities performed by a software development team.* To select the activities carried out by a software development team, as Fig. 2 shows, an analysis between the phases, activities and the roles' scripts of the TSPi® methodology were performed (Humphrey, 2006).

2. *Identify the activities used for the integration of teams.* To achieve this activity, a set of three activities was carried out: (1) search the activities implemented for the integration of teams, using academic resources and on-line resources; (2) classify the activities, focusing on the following criteria: objective to achieve, number of necessary participants, and required time for playing the game; (3) identify the candidate activities, based on three criteria—objective easy to understand, an activity for five to 12 participants, and the time to play the game has to be not more than 60 minutes. Table 2 shows the result of Steps 1 and 2.

   After analyzing the set of candidate activities, the *Rube Goldberg* machine was selected. This activity was designed to motivate teamwork and the resolution of problems with students of all ages (Rube Goldberg, 2016).

3. *Map the activities of the TSPi® methodology and the selected activity.* All activities contained in the TSPi® methodology (Humphrey, 2006) phase were analyzed, based on the performance of the Rube Goldberg machine. As result of the mapping, a set of phases was identified as part of the Rube Goldberg machine and linked to the activities carried out by a development team, such as *(1) planning; (2) design; (3) development (implementation); and (4) tests.*

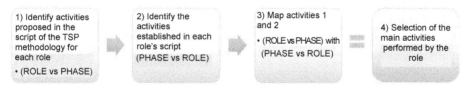

Fig. 2: Steps performed to select the main activities performed by a development team.

Table 2: Shows the main activities identified to integrate a team.

Activity	Goal	Number of Participants	Game Time
Align by influence	To identify a leader of a team in an impartial way	12	60 min
Team work	To show the efficiency in team works	From 5 to 7	30 min
Pulse	To integrate a new team member	N/A	N/A
Situation space	To create relationships among team members	N/A	15 min
Rube Goldberg machine	To reinforce the team work	At least 3	N/A
The race of cars	To show the efficiency in team works	From 5 to 7	20 min
Building a team	To increment the cohesion and identity of team works	8	30 min

N/A: does not apply

4. *Building the virtual environment.* Based on the mapping performed in the previous step, the virtual environment of the Rube Goldberg machine was developed. The Rube Goldberg machine idea consists of starting with a person launching a simple action that carries out a set of chain reactions until the objective or challenge is achieved (Rube Goldberg, 2016). Some examples of challenges are to cook a hamburger, turn-on or turn-off an alarm, or toast a slice of bread among others.

The Rube Goldberg machine requires specific characteristics (Rube Goldberg, 2016), such as (1) it has to be a large structure; (2) it has to be complicated to create; (3) it includes a number of chain reactions, usually very easy tasks; (4) encourages teamwork; and (5) encourages the solution of problems.

The Rube Goldberg machine was selected because it creates a set of phases, such as planning, designing, development (implementation), and testing; phases that can be mapped to the performed activities by a software development team.

In our case, the selected goal of the Rube Goldberg machine is to organize some elements to ring a bell carrying out the necessary chain reactions. To achieve it, the selected elements with physical properties of size and weight as in real life, were a chair; a volleyball ball; an inclined wooden board; a set of 15 dominoes; a bell; and, two additional elements: a soccer ball and an extra domino with extra weight compared to real life. The use of these last elements was to force the activities of planning, designing and testing within the game.

In this way, the game activities reflected the software development activities, and consequently, that each member left their comfort zone. This situation will allow us to perform the analysis to identify the appropriate role for each team member, which is the main motivation in the virtual reality environment.

4.2 Development Methodology for Building the Virtual Reality Environment

For the development of the software tool in this research, the platform created by Valdez et al. (2014) and described in 'development of a prototype of a collaborative virtual environment 3D for the study of the interaction between users; was modified.

Therefore, the five steps followed to develop the virtual reality environment were:

1. *Modeling the objects for the virtual environment.* In this step, two activities were performed: definition of a set of objects for the virtual environment, and modeling of the objects. The result of this step was 3D models, developed with the modeling software tool, Blender™.

2. *Coupling the objects in the virtual environment and adapting the physical properties of the objects.* In this step, two activities were carried out: the attachment of 3D objects into the Unity 3D project and the addition of the physical properties.

   As a result of performing both the activities, the objects of the game scene were obtained (Fig. 3).

3. *Defining the gamification elements in the virtual environment.* In this step, the definition of the gamification elements, that were implemented in the virtual reality environment, was made. The main gamification elements identified were:

   • *Leaderboards*: These gamification elements are useful to show the user's progress or success. Besides, it allows comparing an individual's progress or success with that of the other users.

   • *Points*: These gamification elements are useful to measure the success of an individual on fulfilling an activity.

   • *Badges*: These gamification elements are useful to represent the success or achievement of personal goals and, therefore, to motivate an individual to achieve his/her goals.

   • *Levels*: These gamification elements are useful to provide challenges or to increase the challenge difficulty, depending on an individual's progress.

   • *Progress bars*: These gamification elements are useful to provide an overview of the achieved progress when an individual performs an activity.

   • *Rewards*: These gamification elements are useful to make a gift to an individual in recognition of his/her performed work.

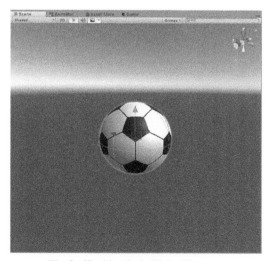

Fig. 3: 3D object in the Unity 3D scene.

- *Scores*: These gamification elements are useful to measure a user performance while executing an activity.
- *Challenges*: These gamification elements are useful to make users apply their knowledge and skills to achieve an activity.
- *Achievements*: These gamification elements are useful to represent the specific goals in a main activity.
- *Feedback*: This gamification element is useful to provide information messages to identify how to perform the activities.
- *Unblocking content*: This gamification element is useful to unblock activities when users achieve goals.

Four gamification elements to be implemented in the virtual environment were selected points, levels, rewards, and feedback. Table 3 presents the details of each gamification element, as well as its mode of use in the virtual reality environment.

4. *Constructing the virtual environment.* This step consisting of development of the virtual environment, entails three activities: (1) check the functioning of the game objects added in Step 1; (2) design the interaction between the user and the gamification elements; and (3) validate the operation between the game object and

Table 3: Selected gamification elements for the tool.

Gamification Element	Functionality	Function	Object in the Virtual Environment
Points	Showing the accumulated points according to the acceptance criteria to build the Rube Goldberg machine. For example, 10 points, if the team uses all the objects to construct the machine.	The accumulated points are shown in the screen with a starred figure.	Puntos Ganados 10
Levels	The actual level is indicated. The game has three levels with different objects to construct the Rube Goldberg machine.	A label in the screen shows the game level.	Nivel: 1 de 3
Rewards	A reward of five extra minutes is given to the participants in case they achieve the next level.	A message with the cumulated time is presented in a label on the screen.	4:15.31
Immediate feedback	Messages with information regarding the objects are presented with some ideas on how to construct the machine.	A message in the center of the screen is presented.	Las fichas de dominó pueden causar buenas reacciones en cadena. ¡Úsalas con cuidado! The Spanish message says: 'Dominoes can cause a nice chain reaction. Use them carefully.'

the gamification elements. The virtual environment developed is shown in described in Section 5.

5. *Proof and validation of the virtual environment.* This step aims to carry out a set of tests that allow validating the virtual reality environment.

5. Developed Software Tool

This section presents screenshots of the virtual reality environment for the identification of work roles. The screenshots have been organized in four groups: (1) start of the application; (2) game scene; (3) displacement of objectives; and (4) proof of the construction made.

1. *Start of application*: This group presents the screenshots to initiate the virtual reality environment. Figure 4 shows the captured user name; Fig. 5 shows the server setup, and Fig. 6, the *avatar* selection.

2. *Game scene*: Figure 7 shows in the scene shown to the team members with previously located objects. The goal, as mentioned,will be to reorganize them in order to ring the bell carrying out the necessary chain reactions.

3. *Displacement of objectives*: Figure 8 presents the screenshots of a moving soccer ball to relocate it.

Fig. 4: Virtual environment - home.

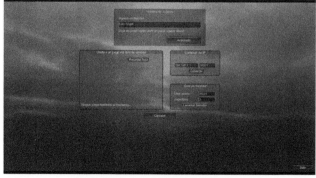

Fig. 5: Virtual environment - start server.

Fig. 6: Virtual environment-*avatar* selection.

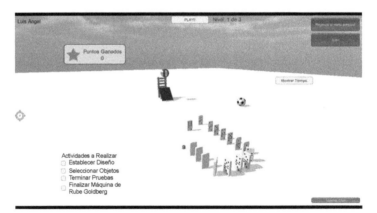

Fig. 7: Virtual environment - soccer ball and extra domino tab.

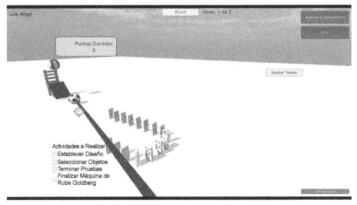

Fig. 8: Virtual environment - displacement of an object: soccer ball.

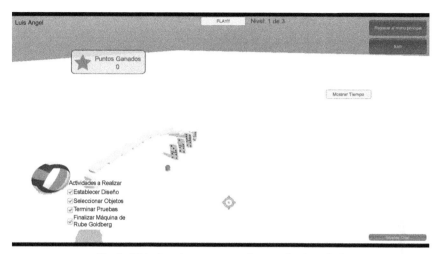

Fig. 9: Virtual environment - proof game - domino tabs 2.

4. *Proof of the construction made*: Figure 9 shows screenshots with the requested objective achievement. On completion of the planning and designing activities for the team members, a construction test must be carried out. The test is performed by pushing the 'play' button, which adds gravity to the playing elements, and therefore, starting the chain reactions.

6. Evaluation and Discussion

A case study was performed in order to evaluate the viability of the virtual environment to integrate high effective teams. The case study aims to identify the roles for a team considering the six roles of TSPi (see *Team Software Process* of Section 2.1).

Thus the virtual reality environment was used by two teams in order to identify and provide the roll for each member. Table 4 shows the teams' features.

The research questions their goal, and the survey question to be answered by user defined for this case study are shown in Table 5.

A summary of the obtained results from the case study is shown in Table 6, in a graphic mode.

The concept of building a Rube Goldberg machine, that is, the general idea of this proposal allows building a set of chain reactions to achieve an objective, the use of a virtual reality to build the Rube Goldberg machine, and facilitate its construction because in real life, one of the main features of this machine is that it has to present a very difficult and complex structure to create. Also, changing physics to objects is not possible in real life.

This paper presents a virtual reality environment that involves three components: (1) the identification of teamwork activities; (2) the study of interactive styles; and (3) the application of gamification elements (dynamics, mechanics, and components). Together, the three elements have the objective of the identification of roles in a software development team that can be assigned with the support of the activities performed in the game.

Table 4: Teams' features.

Team	Type of Team	Number of Members	Features
Team 01	Academic software development team	4 team members (1 woman and 3 men)	Both teams members are students of a software engineering master
Team 02	Academic software development team	4 team members (2 women and 2 men)	

Table 5: Research questions established for the case study.

Research Question	Goal	Survey Question
RQ1. The techniques used by the teams for assigning work roles are adequate?	This question allows exploring if the techniques used by the teams for the assignment of the work roles achieve a good performance in the development of the activities.	• Do you consider that the objective of the activity has been achieved? • Do you consider that the role played by your person was adequate? • Do you consider that the roles played by the other team members were adequate? • Do you consider that the performance of the team was better with the suggested roles of the tool?
RQ2. Does the tool facilitate the identification of the work roles for the members of a development team?	This question allows identifying if the work role assigned to a person is adequate, according to the results of the activities carried out in the tool.	• Do you consider that the tool is adequate to identify the work role? • Do you consider that the role suggestions for the other team members were correct?
RQ3. Is the method to perform the assignment of the role by means of the tool adequate?	This question allows verifying the identification of a person regarding a work role.	• Do you consider that the suggestion of your role was adequate? • Do you think that the roles identified with the tool seem appropriate?
RQ4. Does the tool have a level of usability acceptable to the user?	This question allows validating the tool usability level.	• Is the tool easy to use? • Do you find the design and appearance of the tool attractive?

During the use of the tool, the team members of both teams demonstrate a different attitude regarding their attitude in real life. This indicates that applying a game to perform activities related to the integration of a team exposes people to leave their comfort zone, so that we can observe the natural behavior of the team members, which highlights their abilities: team abilities, knowledge and skills, avoiding a closed environment in which they might feel observed and pressured.

Then, the use of the virtual reality and the gamification elements present an attractive way in which a work profile that best matches each member can be identified.

7. Conclusions and Future Work

The assignment of roles is a key activity prior to software development. Unfortunately most of the times this activity is done in a light way, based on the team's fellowship preferences, which not always result in a correct role assignment.

Table 6: Summary of case study results.

Survey Question	Results
• Do you consider that the objective of the activity has been achieved?	
• Do you consider that the role played by your person was adequate?	
• Do you consider that the role played by the other team members was adequate?	

Table 6 contd. ...

... Table 6 contd.

Survey Question	Results
• Do you consider that the performance of the team was better with the suggested roles of the tool?	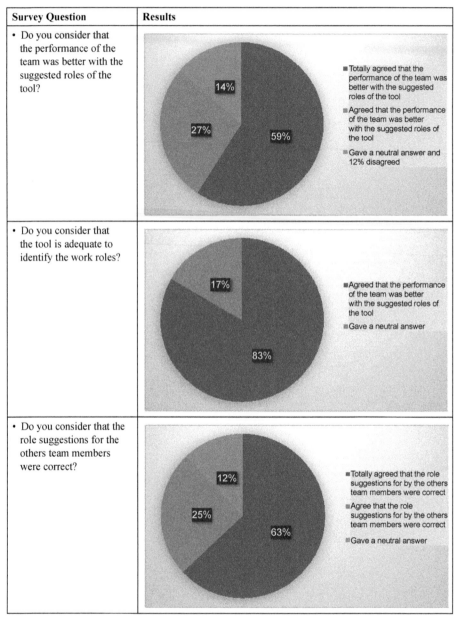
• Do you consider that the tool is adequate to identify the work roles?	
• Do you consider that the role suggestions for the others team members were correct?	

Table 6 contd. ...

... Table 6 contd.

Survey Question	Results
• Do you consider that the suggestion of your role was adequate?	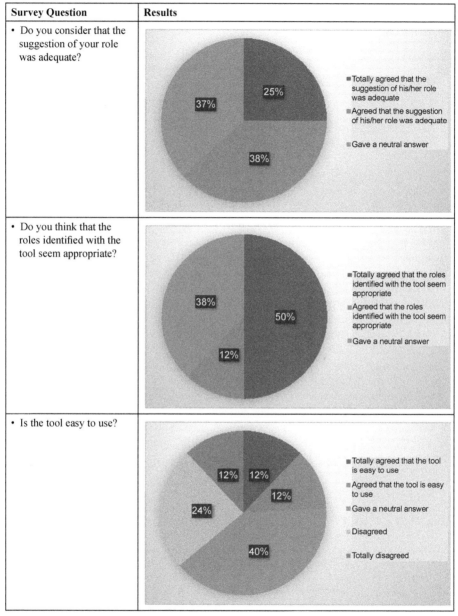
• Do you think that the roles identified with the tool seem appropriate?	
• Is the tool easy to use?	

Table 6 contd. ...

... Table 6 contd.

Survey Question	Results
• Do you find the design and appearance of the tool attractive?	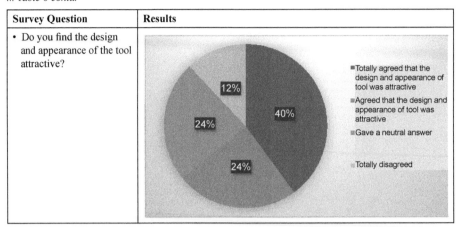

By performing the assignment through virtual reality, we concluded that this activity should be done on the basis of skills and knowledge of each person in order to constitute an effective team.

Besides, according to the case study results, the virtual reality environment showed excellent results regarding the identification of work roles for each team member; the improvement of team performance with the role suggestions and proving attractive for the users. However, it identified an improvement opportunity related to the ease in use. But taking into account all the results, it can be said that virtual reality as a tool for identifying roles in order to integrate highly effective teams achieved its objective. It is important to mention that the main limitation of the use of virtual reality as a tool is that it needs an expert to analyze the team members during the use of the virtual environment.

Based on the case study results and the above-mentioned limitations, as future work, the virtual reality environment could be improved according to the results of the case study: automate the analysis of the role identification, and add more gamification elements to make it more attractive.

References

Abdullah, N.A.S., Rusli, N.I.A. and Ibrahim, M.F. (2014). Mobile Game Size Estimation, 2014 IEEE Conference on Open Systems (ICOS), 42–47.

Ali, Z. and Usman, M. (2017). A framework for game engine selection for gamification and serious games. FTC 2016—Proceedings of Future Technologies Conference (December), 1199–1207, https://doi.org/10.1109/FTC.2016.7821753.

Andreoli, R., Erra, U. and Dipartimento, I. (2005). Interactive 3D Environments by Using Videogame Engines.

Benefield, G.A., Shen, C. and Leavitt, A. (2016). Virtual team networks: How group social capital affects team success in a massively multiplayer online game. *In*: Proceedings of the 19th ACM Conference on Computer-supported Cooperative Work & Social Computing—CSCW, 2016, pp. 677–688, https://doi.org/10.1145/2818048.2819935.

Bozanta, A., Kutlu, B., Nowlan, N. and Shirmohammadi, S. (2016). Effects of serious games on perceived team cohesiveness in a multi-user virtual environment. Computers in Human Behavior, 59: 380–388, https://doi.org/10.1016/j.chb.2016.02.042.

Bricken, M. (1991). Virtual reality learning environments: Potentials and challenges. AcmSiggraph Computer Graphics, 25(3): 178–184.

Burke, B. (2014). Gamify How Gamification Motivates People to do Extraordinary Things, Bibliomotion, Inc.

Christel, M.G., Stevens, S.M., Maher, B.S., Brice, S., Champer, M., Jayapalan, L. and Lomas, D. (2012). Rumble Blocks: Teaching science concepts to young children through a unity game. Proceedings of CGAMES'2012 USA—17th International Conference on Computer Games: AI, Animation, Mobile, Interactive Multimedia, Educational and Serious Games, 162–166, https://doi.org/10.1109/CGames.2012.6314570.

Chou, Y.-K. (2015). Actionable Gamification Beyond Points, Badges, and Leaderboards, Octalysis Media.

Dorling, A. and McCaffery, F. (2012). The gamification of SPICE. 12th International SPICE Conference, SPICE 2012, 290: 295–301.

Ellis, J.B., Luther, K., Bessiere, K. and Kellogg, W.A. (2008). Games for virtual team building. Proceedings of the 7th ACM Conference on Designing Interactive Systems—DIS '08, 295–304, https://doi.org/10.1145/1394445.1394477.

Ellis, S.R. (1995). Virtual Environment and Environment's Instruments, in Simulated and Virtual Realities, pp. 11–51, Taylor & Francis, Inc. Bristol, PA, USA ISBN: 0-7484-0129-6.

Guenaga, M., Eguiluz, A., Rayon, A., Nunez, A. and Quevedo, E. (2014). A serious game to develop and assess teamwork competency. 2014 International Symposium on Computers in Education, SIIE 2014, 183–188, https://doi.org/10.1109/SIIE.2014.7017727.

Häkkinen, P., Bluemink, J., Juntunen, M. and Laakkonen, I. (2012). Multiplayer 3D game in supporting team-building activities in a work organization. Proceedings of the 12th IEEE International Conference on Advanced Learning Technologies, ICALT, 2012, 430–432, https://doi.org/10.1109/ICALT.2012.242.

Hernández, L., Muñoz, M., Mejía, J., Peña, A., Rangel, N. and Torres, C. (2017). A systematic literature review focused on the use of gamification in software engineering teamworks. Revista Ibérica de Sistemas y Tecnologías de la Información, 21(3): 33–50.

Hernández-López, A., Colomo-Palacios, R., García-Crespo, A. and Soto-Acosta, P. (2010). Team Software Process in GSD Teams: A study of new work practices and models. International Journal of Human Capital and Information Technology Professionals, 1(3): 32–53.

Herranz, E., Colomo-Palacios, R. and de AmescuaSeco, A. (2015): Gamiware: A gamification platform for software process improvement. *In*: O'Connor, R., Umay Akkaya, M., Kemaneci, K., Yilmaz, M., Poth, A. and Messnarz, R. (eds.). EuroSPI 2015, CCIS, vol. 543, Springer, Cham, https://doi.org/10.1007/978-3-319-24647-5_11.

Humphrey, W.S. (2006). Introduction to the Team Software Process, Addison-Wesley.

Kapp, K.M. (2012). The Gamification of Learning and Instruction: Game-based Methods and Strategies for Training and Education, first ed., Pfeiffer & Co.

Kim, H., Kang, Y., Shin, S., Kim, I. and Han, S. (2014). Collaborative visualization of a warfare simulation using a commercial game engine. Lecture Notes in Computer Science (Including Subseries Lecture Notes in Artificial Intelligence and Lecture Notes in Bioinformatics), 8526 LNCS (PART 2), 390–401, https://doi.org/10.1007/978-3-319-07464-1_36.

Knutas, A., Ikonen, J., Nikula, U. and Porras, J. (2014). Increasing collaborative communications in a programming course with gamification. Presented at 15th International Conference on Computer Systems and Technologies, http://dx.doi.org/10.1145/2659532.2659620.

Lukosch, H., Nuland, B. van, Ruijven, T. van, Veen, L. van and Verbraeck, A. (2014). Building a Virtual World for Team Work Improvement, pp. 60–68, https://doi.org/10.1007/978-3-319-04954-0_8.

Messaoudi, F., Simon, G. and Ksentini, A. (2016). Dissecting games engines: The case of Unity 3D, Annual Workshop on Network and Systems Support for Games, 2016, Janua. https://doi.org/10.1109/NetGames.2015.7382990.

Muñoz, M., Mejia, J., Peña, A. and Rangel, N. (2016). Establishing effective software development teams: an exploratory model. Communications in Computer and Information Science, 425: 13–24, https://doi.org/10.1007/978-3-662-43896-1.

Muñoz, M., Hernández, L., Mejia, J., Peña, A., Rangel, N., Torres, C. and Sauberer, G. (2017). A Model to Integrate Highly Effective Teams for Software Development, System, Software and Services

Process Improvement, Springer International Publishing, AG, J. Stolfa et al. (eds.). EuroSPI 2017, CCIS 748, pp. 613–626, DOI: 10.1007/978-3-319-64218-5_51.

Muñoz, M., Peña, A., Mejía, J., Gasca-Hurtado, G.P., Gómez-Álvarez, M.C. and Hernández, L. (2018). Applying gamification elements to build teams for software development. IET Software, DOI: 10.1049/iet-sen.2018.5088 IET Digital Library.

Noh, S.S., Hong, S.D. and Park, J.W. (2006). Using a game engine technique to produce 3D entertainment contents. Proceedings—16th International Conference on Artificial Reality and Telexistence - Workshops, ICAT 2006, 246–251, https://doi.org/10.1109/ICAT.2006.139.

Pavkov, S., Franković, I. and Hoić-Božić, N. (2017). Comparison of game engines for serious games. 2017 40th International Convention on Information and Communication Technology, Electronics and Microelectronics, MIPRO 2017—Proceedings, 728–733, https://doi.org/10.23919/MIPRO.2017.7973518.

Peña, A. and Jiménez, E. (2012). Virtual environments for effective training. Revista Colombiana de Computación—RCC, Universidad Autónoma de Bucaramanga, 13(1): 45–58, ISSN: 1657-2831.

Polančec, D. and Mekterović, I. (2017). Developing MOBA games using the Unity game engine. Information and Communication Technology, Electronics and Microelectronics (MIPRO), 2017, 40th International Convention, 1510–1515, https://doi.org/https://doi.org/10.23919/MIPRO.2017.7973661.

Rangel, N., Torres, C., Peña, A., Muñoz, M., Mejia, J. and Hernández, L. (2017b). Team members' interactive styles involved in the software development process. pp. 675–685. *In*: Stolfa, J., Stolfa, S., O'Connor, R.V. and Messnarz, R. (eds.). Systems, Software and Services Process Improvement: 24th European Conference, EuroSPI 2017, Ostrava, Czech Republic, September 6–8, 2017, Proceedings, Cham: Springer International Publishing, https://doi.org/10.1007/978-3-319-64218-5_56.

RibesIñesta, E. (1990). El problema de las diferencias individuales: un análisis conceptual de la personalidad. pp. 231–253. *In*: Problemas conceptuales en el análisis del comportamiento, México: Trillas.

RibesIñesta, E. (2009). La personalidad como organización de los estilos interactivos. Revista Mexicana de Psicología, 26(2): 145–161.

Rube Goldberg, I. (2016). Rube Goldberg Machine Contest® 2016 Official Rule Book, July 2015, 1–19.

Schroeder, R. (2006). Being there together and the future of connected presence. Presence: Teleoperators and Virtual Environments, 15(4): 438–454.

Schweiger, N., Meusburger, K., Hlavacs, H. and Sprung, M. (2014). Jumru 5s—A game engine for serious games. pp. 107–118. *In*: Ma, M., Oliveira, M.F. and BaalsrudHauge, J. (eds.). Serious Games Development and Applications: 5th International Conference, SGDA 2014, Berlin, Germany, October 9–10, 2014, Proceedings, Cham: Springer International Publishing, https://doi.org/10.1007/978-3-319-11623-5_10.

Söbke, H. and Streicher, A. (2016). Serious games architectures and engines. pp. 148–173. *In*: Dörner, R., Göbel, S., Kickmeier-Rust, M., Masuch, M. and Zweig, K. (eds.). Entertainment Computing and Serious Games: International GI-Dagstuhl Seminar 15283, Dagstuhl Castle, Germany, July 5–10, 2015, Revised Selected Papers, Cham: Springer International Publishing, https://doi.org/10.1007/978-3-319-46152-6_7.

Spante, M., Heldal, I., Steed, A., Axelsson, A. and Schroeder, R. (2003). Strangers and friends in networked immersive environments: Virtual spaces for future living. Proceeding of Home Oriented Informatics and Telematics (HOIT).

Steffens, F., Marczak, S., Filho, F.F., Treude, C., Singer, L., Redmiles, D. and Al-ani, B. (2015). Using Gamification as a Collaboration Motivator for Software Development Teams: A Preliminary Framework, 48–55.

Uskov, A. and Sekar, B. (2014). Serious games, gamification and game engines to support framework activities in engineering: Case studies, analysis, classifications and outcomes. IEEE International Conference on Electro Information Technology, 618–623, https://doi.org/10.1109/EIT.2014.6871836.

Valdez Gómez, E. and Peña, A. (2014). Desarrollo de un prototipo de entorno virtual colaborativo 3D para el estudio de la interacción entre usuarios. Universidad de Guadalajara.

Vasudevamurt, V.B. and Uskov, A. (2015). Serious game engines: Analysis and applications. IEEE International Conference on Electro Information Technology, 2015 June, 440–445, https://doi.org/10.1109/EIT.2015.7293381.

Volkswagen. (2010). The Speed Camera Lottery, retrieved from http://www.thefuntheory.com/speed-camera-lottery-0.

Wendel, V., Gutjahr, M. and Battenberg, P. (2013). Designing a Collaborative Serious Game for Team Building Using Minecraft, pp. 569–578, October, retrieved from http://search. proquest.com/openview/f11c9d6e0d896ca4a090c285fc678ce7/1?pq-origsite=gscholar.

Westhoven, M. and Alexander, T. (2015). Towards a structured selection of game engines for virtual environments. pp. 142–152. *In*: Shumaker, R. and Lackey, S. (eds.). Virtual, Augmented and Mixed Reality: 7th International Conference, VAMR 2015, Held as Part of HCI International 2015, Los Angeles, CA, USA, August 2–7, 2015, Proceedings, Cham: Springer International Publishing, https://doi.org/10.1007/978-3-319-21067-4_16.

Werbach, K. and Hunter, D. (2012). For the Win: How Game Thinking Can Revolutionize Your Business, Wharton Digital Press.

Chapter **8**

Integrating Virtual Reality into Learning Objects-based Courses

Jaime Muñoz-Arteaga[1,*] and *Héctor Cardona Reyes*[2]

1. Introduction

Nowadays, virtual reality has the advantages to include recreating different e-learning environments and situations from real life or hypothetical scenarios to learn. Given the heterogeneity of platforms, standards, and instructional design, it becomes difficult to integrate virtual reality in online courses. In addition, learning objects are considered as educational resources that can be employed in technology-support learning. This work proposes a model of learning objects integrating virtual reality mechanism in order to offer a better support for the teaching-learning process. In the proposed model, the learning object represents pieces of knowledge to put together, so as to design and develop online courses with several components with different granularity levels. Current proposal is applied in a case study describing the analysis and the design of a course composed by learning objects with virtual reality.

A learning object (LO) is defined as a self-standing, reusable, discrete piece of content broken down into smaller chunks that can be reused in any environment in order to meet an instructional objective (Muñoz-Arteaga et al., 2007). They can be presented in many forms, such as web pages, PDF documents, video, audio, animation, and 3D representations (Abud, 2012). Nowadays, virtual reality has the advantages to recreate different e-learning environments and ludic situations from real life or hypothetical scenarios to learn. Given the heterogeneity of platforms, standards and instructional design it becomes difficult to integrate virtual reality in online courses (Martinez et al., 2006).

---

[1] Universidad Autónoma de Aguascalientes, Mexico.
[2] CIMAT Zacatecas, Mexico.
* Corresponding author: jaime.munoz@edu.uaa.mx

Learning objects have been developed in order to technologically and pedagogically support the virtual education. Moreover, such contents can be used under any educational circumstances, either in training or in distribution of the knowledge as required in classroom lessons, in staff training in the industry, or in self-learning process, and so on. However, in the e-learning area, there are several works that offer few options for access to educational resources with virtual reality (Abud, 2012; Garcia et al., 2006). In general, teachers do not have access to courses composed of learning objects with virtual reality. Some other difficulties are identified as follows:

- Learning objects with virtual reality development require, since the early phases, different perspectives of stakeholders, such as teachers, psychologists, programmers, students, analysts, etc. (Muñoz-Arteaga et al., 2007).
- What is the instructional design for courses related to learning objects with virtual reality? (Garcia et al., 2006).
- It lacks criteria to evaluate the quality of this kind of learning of objects (Muñoz-Arteaga et al., 2007).
- The learning objects with virtual reality need to be portable and reusable across multiple platforms (Abud, 2012).
- The navigation style and the type of interactions are defined in terms of course that house the learning objects with virtual reality (Martinez et al., 2006).
- Teachers do not have access to training programs on the use of new technologies to apply inside and outside the classroom.

In order to mitigate difficulties related to designing educational content, the current work proposes a set of models for design courses in terms of learning objects with virtual reality. In this work, we will not be focusing on the strictly pedagogical aspects, but on some design issues perceived as difficult in the adoption of educational content. To this end, Section 2 presents a learning object model with virtual reality. Sections 3 and 4 propose respectively a structure and architecture for a course composed of learning objects with virtual reality coming from different repositories. Section 5 proposes a methodology to develop learning objects. Section 6 presents in detail a case study in order to show the feasibility of the current proposal.

2. Learning Object Model with Virtual Reality

Virtual reality headsets or multi-projected environments can be used to generate realistic images, sounds and other sensations that simulate a user's physical presence in a virtual environment. The virtual reality market is constantly growing; new developments occur almost every month, increasingly sophisticated, and offer more possibilities to the player. Nowadays virtual reality is being used in multiple platforms. Thus, the user can explore and interact with several academic resources offered in a learning environment through virtual reality. The virtual reality mechanisms can be considered in some elements of a learning object, particularly in theoretical and practical knowledge (Fig. 1). With this model, the students can apply different alternatives to learn, for example, it is possible to put in practice the theoretical knowledge by playing with the virtual reality content in every learning object.

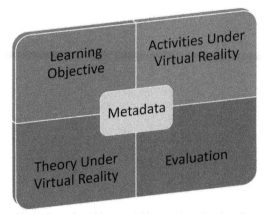

Fig. 1: Learning object model integrating virtual reality.

The components of a learning object are described in the model of Fig. 1 as follows:

- The learning objective establishes and articulates the academic expectations.
- Theory under virtual reality specifies the foundation of educational content in a certain subject of knowledge.
- The activities offer interactive virtual reality scenarios to put in practice the theoretical foundation.
- The evaluation under virtual reality offers interactive validation related to theoretical and practical knowledge.
- The metadata element is an auto description concerning the learning object.

An example of theoretical content is represented in the Fig. 2 where a skull is shown in virtual reality, representing different perspectives and elements of a human eye in terms of a learning object.

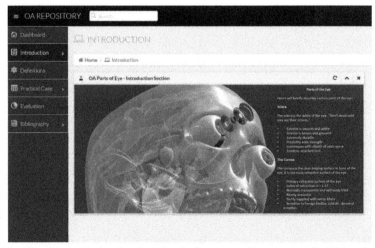

Fig. 2: A learning object with virtual reality related to human eye (*Source*: e-learning University Medical Center, 2019).

Furthermore, the concept of reusable learning objects has evolved due to the need to standardize and reuse online learning material. In order to become reusable, a learning object must include information about its contents in the form of metadata. This metadata allows the object to be indexed, making it easy to store and to retrieve from a repository. The Learning Object Metadata (LOM) according to IEEE standard establishes which the kind of information that can be stored to assure interoperability between learning objects repositories (Advanced Distributed Learning, 2019). The use of standard SCORM in a learning object with virtual reality offers some economics and pedagogical advantages: economic advantages (e.g., create once, use several times) as well as pedagogical advantages (e.g., high-quality interactive multimedia easily available for courses in individualized learning).

3. Learning Object Course with Virtual Reality

Learning objects are considered as educational resources that can be employed in technology-support learning. They are digital pieces of knowledge meant to be put together in order to conform to a course. The authors propose a UML model in Fig. 3 in order to describe the components of a course, which can be composed of several modules or units, with every module have a set of subjects, and every subject can be specified by one learning object with virtual reality. The model of Fig. 3 specifies a certain number of attributes and methods per each object learning; the methods can be meant to obtain the respective attributes, such as metadata, evaluation, learning objective, theoretical and practical knowledge.

In general, a learning objects with virtual reality can be produced by teachers and technologists from educative institutions to extend digital educative content to various devices, such reality headsets or multi-projected environments. Here a course online is accessible, using a virtual learning environment inserted in a learning object. The granularity and the instructional design of educational content for modules, subjects, and learning objects are defined by the teacher. Next, a designer and a teacher can define the interactive learning activities throughout an environment with virtual reality.

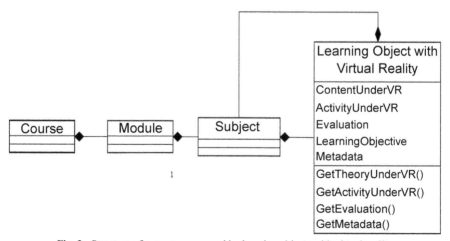

Fig. 3: Structure of a course composed by learning objects with virtual reality.

4. Architectural Model for Course with Learning Objects

This work proposes an architectural model composed of three layers: graphical user interface of learning object with virtual reality, learning management system, and repositories management (Fig. 3). The management of layer repositories is composed by distributed repositories offering services (save, update, researcher, order, etc.) to access learning objects, which can be used in online course. In the second layer, the teachers can use a learning management system to define the structure of online course with different modules composed by web services. Finally, the layer of user interface allow the user to access learning objects from an online course using, in general, virtual reality headsets or multi-projected environments.

Nowadays, a large number of universities are producing online courses in terms of learning objects and saving these objects in their repositories, which, in general, support several queries with different criteria, thanks to the information saved in the metadata. There is a plethora of virtual learning environments or learning management systems. The needed requirements for an interaction system in the learning domain should include: first, the facilities to interact during and after the lecture; second, an open architecture which includes the possibility of allowing extensions; and third, the system must be scalable (i.e., the system should be able to manage a single course or a whole organization). We are going to use these requirements as the base for comparison between learning management systems (LMSs).

Open-source learning management systems (Moodle (Dougiamas, 2019) and Dokeos (Dokeos, 2019)) can be extended in a such way that a teacher can find and select learning objects in order to develop online courses (Fig. 4). Thanks to the connection of repositories, it is possible to reuse the academic contents and offer online courses in terms of learning objects. In addition, all the learning objects in the repositories should be filtered by researchers asking and getting learning objects with virtual reality. Then, the students can use these learning objects in a course under different platforms, such as lap, tablets, lens, and mobiles devices.

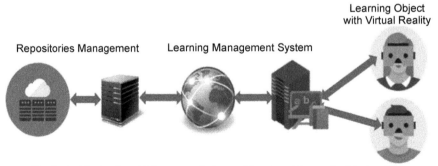

Learning Object
with Virtual Reality

Repositories Management   Learning Management System

Fig. 4: Architectural model for courses in terms of learning objects with virtual reality.

5. Methodology to Develop Learning Objects with Virtual Reality

Current work uses the methodology MEDOA (*Metodología para el Desarrollo de Objetos de Aprendizaje*) for the development of learning objects with virtual reality. This methodology is shown in Fig. 5. It has seven phases that include planning, analysis, design, implementation, validation, deployment and maintenance. The spiral phases (analysis, design, implementation and validation) are most time-consuming to gather information, notably the educational virtual reality content and the development process. In fact, the MEDOA methodology (Alonso et al., 2012) is a mixed process model coming from waterfall and spiral software process (Sommervile, 2015). The start of the strategy considers the planning phase and then the spiral style iteration that involves the four phases of the methodology.

As a matter of fact, MEDOA methodology can be applied in an effective way to develop the structure of a LO with virtual reality. First of all, it is necessary define the learning objective, which is analyzed, designed, and implemented with a specific content that must respond to it. Subsequently, the activities are analyzed, designed, and implemented, which reinforces the content related to virtual reality mechanisms, such as navigation, animation, and simulation. Therefore, the need to finalize the development of the content is necessary from which the activities are built. Finally, the same procedure is applied with evaluation. It is necessary to go through the same phases to generate them under the defined content and activities.

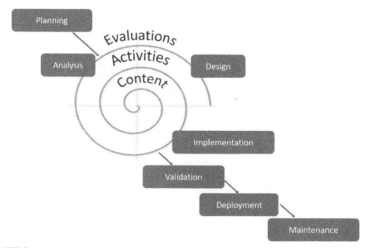

Fig. 5: MEDOA, a methodology for development of learning objects with virtual reality (Alonso et al., 2012).

Once the course of the four phases in its three iterations is completed, then it is again a cascade strategy, but this time it is to execute the validation and implementation phases. Maintenance, although within this process, is an additional phase that is included within MEDOA to promote a review of the learning object developed after two years of its creation or to make some changes in it if some problem is detected in its operation, or its design or its content. This phase has been used at times to standardize learning objects that have been developed under other methodologies.

Another aspect to take into account in this methodology is that it has been implemented computationally and as a result, the methodological computational case tool (Bartocci and Lió, 2019) can be used to document the entire development related to information on LO with virtual information.

6. Case Study

This section presents the specifications of a course in biochemical engineering taking into account several learning objects with virtual reality. This course is specified according to the models proposed in previous sections. The course is unregistered under the Moodle platform of *Universidad Autónoma de Aguascalientes* in México (Universidad Autónoma de Aguascalientes, 2019). In general, a teacher can select some learning objects by requesting for some repositories; for this, it is necessary to launch a query on learning object using key words, followed by the research service start in the local repository. The service can take the result and display all information in a unique list of learning objects.

6.1 *A Cellular and Molecular Biology Course*

The cellular and molecular biology course has several components distributed under a tree structure (Fig. 6). This course and the physical chemistry course are part of the Bachelor of Science Biochemical Engineering course (Fig. 6). Cellular and molecular biology has two modules: cellular analysis and molecular analysis modules. The first module has the lipid and protein learning objects with virtual reality and the second one has the RNA and biomolecule learning objects with virtual reality.

The user driven-interaction characterizes the course in this case study (Fig. 6). This means that a user always has the option to choose any module and any learning object; for example, when a student selects the module for molecular analysis, then a list of learning objects with virtual reality is displayed for ready use. In addition, this case study takes into account several repositories of three Mexican universities: the UV (*Universidad Veracruzana*), UAA (*Universidad Autónoma de Aguascalientes*) and UPA (*Universidad Politécnica de Aguascalientes*). One of the challenges in teaching biochemistry is to facilitate the selection of digital content; then every learning object

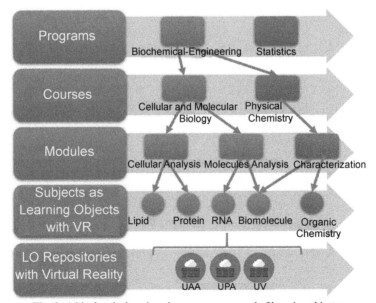

Fig. 6: A biochemical engineering course composed of learning objects.

is designed with highly interactive, interoperable, educational content for carrying out the simulation of biochemical experiments (Martinez, 2006; Harvard University, 2019). These solutions involve the definition of the learning objects in a similar way in a virtual learning environment. Finally, it is possible to use the learning object biomolecule in two different courses (Fig. 5). This reutilization of the learning object is feasible in similar subjects of different courses and can be carried out under the criteria of the main teacher of the course.

6.2 *Management of Learning Object Repository*

The next figure presents an image of repository of learning objects with virtual reality of UAA. This image displays a set of learning objects in the domain of medicine; in particular, a learning object is shown as a theoretical content related to cardiac muscle tissues (Muñoz-Arteaga, 2007). The repository offers the search and selects services of learning objects under different criteria.

Thanks to the use of learning objects in a repository that it has become possible to use them in different courses. The management of a repository offers a series of services, such as display, search, and updated content. Thus a teacher can develop a course in a transparent way by searching and selecting learning objects from different repositories. Several institutions use LMS (Learning Management System) to develop their courses with learning objects in various domains, such as data structure, programming, software engineering, health, etc. (Dougiamas, 2019; Dokeos, 2019; Harvard University, 2019; Universidad Autónoma de Aguascalientes, 2019). Due to the repository it is possible to use heterogeneous learning objects under LMS platforms (Muñoz-Arteaga, 2007), OS platforms, and educational content. Integration of virtual reality in the learning object offers a certain guarantee with portable devices, such as mobile devices, handhelds, laptops, etc.

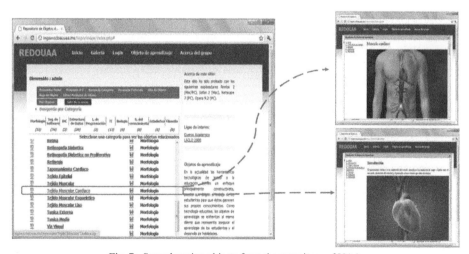

Fig. 7: Some learning objects from the repository of UAA.

6.3 Cellular Analysis Module

The module of cellular analysis of the course in this case study has lipid and protein as learning objects with virtual reality. These learning objects are distributed and displayed in a biochemistry virtual laboratory (Fig. 8). Note that the access to every learning object is not restricted to sequential access; that is, the student has the opportunity to choose and access the theory part first or after the practical part, or vice versa. Consequently, the student can decide on the order to study and this freedom of choice favors the individualization of the learning process. In addition, thanks to the identification of modules, the student can establish relationships between cellular and molecular biology in a biochemistry course.

In addition, the virtual laboratory of this course (Fig. 8) displays information about the progress in a student's activity in every learning object virtual reality.

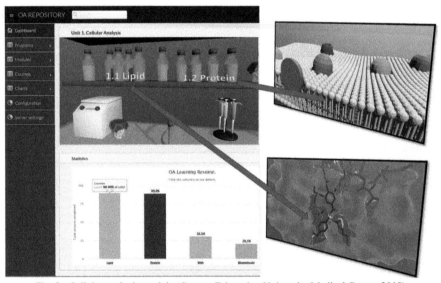

Fig. 8: Cellular analysis module (*Source*: E-learning University Medical Center, 2019).

6.4 Ribonucleic Acid Molecule as Learning Object with Virtual Reality

The molecule ribonucleic acid (RNA) is an important molecule to study in terms of a learning object. The learning objective proposes to define the RNA, compare with other molecules, and put in practice the knowledge of RNA. Next sub-sections describe the rest of the components in the current learning object.

Theoretical content defines ribonucleic acid (RNA) as one of the three major biological macromolecules that are essential for all known forms of life (along with DNA and proteins). Figure 9 presents a brief history about the evolution of RNA research since the 1950s.

Activities content: The LO can be designed to support exploration and investigation of ICT ideas and help in conceptual understanding (Harvard University, 2019). Thus, LOs allow the end users the ability to acquire digital skills.

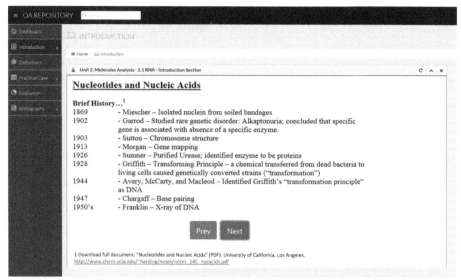

Fig. 9: Theory of RNA learning object (*Source*: University of California, 2019).

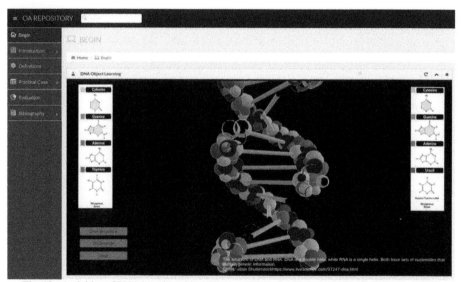

Fig. 10: Activities of RNA learning object (*Source*: E-learning University Medical Center, 2019).

The virtual reality mechanism is a kind of simulation explaining that the RNA acts as a messenger between DNA and the protein synthesis complexes, known as ribosomes, forms vital portions of ribosomes, and acts as an essential carrier molecule for amino acids to be used in protein synthesis (*Source*: E-learning University Medical Center, 2019).

Evaluation content offers an interactive validation of theoretical and practical aspects of RNA learning object class as represented in the window Fig. 11. According

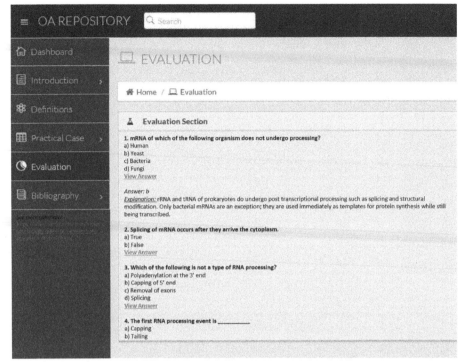

Fig. 11: Evaluation of RNA learning object where the option C is the right answer to question 1 (Source: Manish, 2019): Most newly synthesized RNAs must be modified in various ways to the converted to their functional forms. Bacterial mRNAs are an exception; they are used immediately as templates for protein synthesis while still being transcribed.

to the window, a student can carry out an auto evaluation concerning the RNA subject. Then, it is possible to answer the questions under multiple options; moreover, every user option is possible to compare versus a right answer.

7. Conclusion

The current work proposes a model-centered approach as an alternative for design courses in terms of learning objects with virtual reality. This work, does not focus on the strictly pedagogical aspects, but on some design issues, perceived particularly in integrating virtual reality into learning-objects-based courses. In order to integrate several models at different abstraction levels where several modules conform to a course, every module can be composed of a set of subjects, and every subject can be represented by one learning object with virtual reality.

In addition, a learning object model is proposed where the virtual reality mechanisms can be considered, in particular for theoretical and practical knowledge. Thus it is possible to offer much more different alternatives in learning, for example, it is possible for a student to put in practice his theoretical knowledge throughout the virtual reality mechanisms integrated in every learning object.

Finally, there are several aspects of future work, such as, an extension of standard SCORM to specify an exhaustive metadata of learning objects with virtual reality. It is necessary to define mechanisms to evaluate the quality of this kind of learning objects, taking into account the context, such as the user's profile.

References

Abud, F.A. (2012). Modelo de Objetos de Aprendizaje con Realidad Aumentada. Revista Internacional de la Educación en Ingeniería, 5(1): 1–7, ISSN 1940-1116.

Advanced Distributed Learning. (May 1, 2019). Shareable Content Object Reference Model (SCORM), retrieved from https://adlnet.gov/scorm.

Alonso, M.D.L.A., Castillo, I., Pozas, M., Curiel, A. and Trejo, L. (2012). Estandarizando los Objetos de Aprendizaje con MEDOA. Conferencias LACLO.

Bartocci, E. and Lió, P. (May 1, 2019). The methodological computational case tool, retrieved from https://journals.plos.org/ploscompbiol/article?id=10.1371/journal.pcbi.1004591.

Dokeos. (May 1, 2019). Retrieved from https://www.dokeos.com/.

Dougiamas, M. (May 1, 2019). Moodle, retrieved from: http://moodle.org/.

E-learning University Medical Center. (April 19, 2019). Retrieved from https://ske°tchfab.com/ eLearningUMCG.

Garcia, M.A., Edwards, A., Gutierrez, J. and Acosta, R. (2006). Virtual reality learning objects of molecular structures. pp. 124–128. *In*: International Conference on Dublin Core and Metadata Applications.

Harvard University. (May 1, 2019). Principles of Biochemistry Course, retrieved from https://courses.my.harvard.edu/psp/courses/EMPLOYEE/EMPL/h/?tab=HU_CLASS_SEARCH.

Manish. (April 19, 2019). RNA is Processed in Several Ways: Questions and Answers, retrieved from https://www.sanfoundry.com/molecular-biology-questions-answers-rna-processed-several-ways/.

Martinez, I., Montero, P., Sierra, J.L. and Fernandez, B. (2006). Production and Deployment of Educational Videogames as Assessable Learning Objects. Springer-Verlag.

Muñoz-Arteaga, J., Alvarez, F. and Chan, M. (2007). Tecnología de objetos de Aprendizaje Editorial Universidad Autónoma de Aguascalientes & UdG Virtual, ISBN: 970-728-065-4.

Muñoz-Arteaga, J., Ochoa, X., Calvillo, M.E. and Parra, G. (2007). Integración de REDOUAA a la Federación Latinoamericana de Repositorios de Objetos de Aprendizaje. LACLO 2007, Santiago de Chile.

Sommerville, I. (2015). Ingeniería de Software, tenth ed., Addison Wesley, 10. Edición.

Universidad Autónoma de Aguascalientes. (May 1, 2019). Biochemical Engineering Courses, retrieved from: https://www.uaa.mx/portal/.

University of California, Los Angeles. (April 19, 2019). Nucleotides and Nucleic Acids, retrieved from http://www.chem.ucla.edu/~harding/notes/notes_14C_nucacids.pdf.

Chapter **9**

Virtual Simulation of Road Traffic Based on Multi-agent Systems

Hector Rafael Orozco Aguirre,* Maricela Quintana Lopez,
Saul Lazcano Salas and *Victor Manuel Landassuri Moreno*

1. Introduction

Before the formulation and implementation of the existing norms and regulations for road traffic and pedestrian behavior (Diaz, 2018), people used common sense to move from one location to another. That is to say, in a very simple and unprocessed way, drivers and pedestrians had to look at the environment to keep moving while avoiding a collision or accident. Over the years, several unwritten rules for drivers were accepted, such as the first one to arrive at an intersection had to be the first one to cross it. However, rules and regulations for urban road traffic had to be established alongwith inclusion the use of traffic lights and signs to give the way to some drivers and/or pedestrians and deny the same to others (Khakimov, 2018).

Nowadays, the number of transport vehicles that ply on roads is increasing and although there are rules, regulations, signs and traffic lights, the problems of road traffic and poor pedestrian behavior continue to constantly worsen. Large cities are facing a difficult situation of overcrowding, with a constant increase in the number of vehicles, which lead to daily vehicular traffic congestion. This problem creates difficulties and wastage of time to be taken to move from one place to another. From this perspective, even if certain measures are taken, such as regulating the speed limit on the roads, it is not possible to attend and solve satisfactorily the problems of poor circulation and traffic congestion (Papageorgiou et al., 2003).

On roads, avenues, highways, road junctions, crossings or intersections, vehicular traffic jams are caused due to various reasons, such the bad conditions of the streets, poor planning of the routes, weather, inadequate synchronization of traffic lights,

Autonomous University of Mexico State, Mexico.
* Corresponding author: hrorozcoa@uaemex.mx

lack of road safety education shown by both drivers and pedestrians. Sometimes bad measures aggravate problems and lead to serious accidents that have result in the loss of a considerable number of human lives, or material damage, or affectations in the surrounding environment. With regard to pedestrians, it is dangerous to cross a road lacking access areas (bridges or tunnels), or traffic lights. Thus poor planning, location and distribution of the infrastructure have proven, in practice, to have an impact on the increase of road traffic jams or accident rates.

In recent decades, there has been a geographical and population growth in large cities and metropolitan areas because they are the main centers of economic activities, causing overcrowding in such areas (Mandhare et al., 2018). Concentration of people requires transportation to move from one point to another, leading to an increase in the vehicle index and the need for expansion and improvement of the road infrastructure. Many of the main roads in these areas were not planned for future use at the time of construction; factors, such as geographic and population growth to face the demand of vehicular flow was not taken into consideration. Traffic congestion on main road intersections of cities and metropolitan areas is caused by different factors, such as poor study and planning of road infrastructure, inappropriate traffic light synchronization and lack of vial education.

Because the population and vehicular growth are increasing, in peak hours there is an oversaturation of vehicles on the main roads. This type of road congestion leads to long transportation periods, cause the drivers to lead a stressful life, face the risk of being involved in a road accident, live in an environment plagued by pollution and high noise rates. Traffic congestion produced at a given time and place is the result of a series of individual decisions made by each user of the road infrastructure. Each user decides how and when to travel through what is considered the best path to reach a destination. A decision can be based on criteria such as the cost, time, safety and comfort, as well as the type of transport to be used.

It is possible to avoid or reduce as much as possible the congestion at the intersections and main roads by analyzing the factors that cause it. A convenient approach is the improvement of urban infrastructure by using intelligent traffic lights, capable of giving priority to the most saturated roads, reducing the vehicle load and avoiding possible accidents and traffic jams or by maintaining, upgrading or renewing road infrastructure and pedestrians facilities, such as bridges, overpasses, tunnels, pedestrian crossings among others that permit, in case of traffic jams, possible alternative routes. Virtual modeling allows representation of the common factors and conditions on the road flow along the day. In order to suggest which changes in infrastructure are more convenient, the simulation of the built models offers a path to validate each possible improvement strategy that leads to speeding up of road traffic.

In this chapter, the main factors that cause vehicular congestion in urban areas are considered (Barceló, 2010), through analysis, modeling, and simulation of a real road scenario using the 'AnyLogic Platform' (AnyLogic, AnyLogic Platform, n.d.), which allows to model and simulate a multi-agent system. In order to do this, vehicles, traffic lights, roads, among others, are modeled as agents, making the vehicles capable of taking decisions on going in any direction. With these agents, a case study as a virtual model is carried out to understand why vehicular traffic originates in order to test alternatives or strategies that enable a better control and reduction (Ferrara et al., 2018) for the known Puerto de Chivos road crossing, which is located between the boundaries of the municipalities of Atizapan de Zaragoza and Nicolas Romero in Mexico state, Mexico.

From the evaluation given, it is possible to defend which strategy allows a higher service rate, improving mobility by reducing vehicle queues and travel time (Coclite et al., 2005). These strategies contemplate only experimentation with the synchronization of traffic lights and management during peak hours of traffic congestion, leaving aside others, such as contemplating accidents and improvements in the infrastructure by creating over- or under-passes. This simulation of strategies is to demonstrate their effectiveness and relevance, so that if they were implemented in real life, they help to minimize the effects and problems caused by vehicular traffic.

2. Related Work

3D simulations result in a very attractive way to visualize data and general conditions of traffic simulations as is shown in (Cavallo et al., 2018). The authors constructed a 3D model for trucks and used it in critical lane segments as tunnels, bridges and others in order to show if a problem could occur. They used laser sensors to obtain physical parameters from trucks and use them in their proposed model.

Another work about 3D simulation is presented by Wan and Tang (2003), where they proposed a 3D virtual-reality agent-based model for traffic simulations. Each agent had his own particular behavior and decisions in simulation scenarios. Their simulation model considered original and final locations for each agent/vehicle and total traffic situation as the sum of individual actions of all agents. The proposed model took into account variables as speed, acceptable stopping distance, safe distance between vehicles and others. This work was one of the first 3D simulation scenarios including the individual behavior of the agents.

In order to identify potential risks in new or actual roads, Chen and Wel (2011) proposed a 3D virtual-simulation traffic system that considered physical conditions of the proposed/existing road, pedestrian behavior, and different vehicle types. Their simulations can identify potential road safety, hazards in planned roads, thereby allowing improvement in the original design in order to obtain an improvement safety project. Hongke et al. (2007) used a 3D virtual reality model to simulate different possible scenarios, like traffic jam, fire, normal traffic conditions, accident state and others into a highway tunnel in order to validate action strategies in each of the scenarios proposed.

One important study using 3D simulations is presented by Nemec et al. (2014). They used a 3D virtual car simulation as a tool in drive training and teaching areas. The 3D virtual car simulator used real data and scenarios in order to be more realistic and a useful tool, allowing to check the impact of decisions of driving in the real world. Yu et al. (2013b) proposed a 3D intelligent vehicle model in the virtual-reality scenario. Their proposed model considered the human agent, vehicle agent and environment agent to build a multi-agent model able to simulate complex and hypothetical traffic scenarios.

An important point in simulations is the interaction between agents and the environment in order to obtain more confident results. Haubrich et al. (2014) considered a particular interaction between vehicles and environment into simulations and added road network logics as an invisible layer integrated into virtual scenarios. Road network logics can consider the conditions, like parking zones, low speed areas, tunnels, and street intersections, for example. In this way, another important work was proposed by Berg et al. (2009) in which they reconstructed traffic simulations from data obtained by traffic sensors and used them to construct car models in their proposed scenarios. This work considered a multi-robot planning-problem logic to model behavior of vehicles, adding spatial and temporal constraints in an open-system model trying to search a trajectory that minimizes lane and speed changes.

Tian and Akashi (2013) combined artificial vision algorithms to determine traffic density and type of vehicles involved with using monitoring cameras in order to feed a virtual-reality scenario to simulate traffic conditions, avoiding new sensor networks to obtain traffic data to feed simulation scenarios. Another work in this way by Tang et al. (2017) presented automatic vehicle detection and type recognition model that can be used in a virtual-reality traffic-simulation model to feed traffic density and vehicle type. The proposed model combined artificial vision algorithms with 97 per cent positive detection and 91 per cent positive vehicle recognition while maintaining a fast processing time, allowing implementation of a real-time virtual-simulation system.

Bellini et al. (2018) used a few sensors to feed a mathematical model to simulate traffic density in a whole city area. The proposed mathematical model allowed estimation of real-traffic conditions in city segments without sensors with high accuracy. Their simulations showed that the proposed model can be extended to a macro scenario by adjusting certain parameters. A useful system model to traffic simulation is presented by Yu et al. (2013a), who use a multi-agent platform in a virtual-reality intelligent-simulation system of vehicles to model interaction between vehicles agents and ambient agents. The proposed model can be scalable in order to consider other agents that affect the traffic density in micro or macro scenarios.

Another work that considers the interaction between different kinds of agents is presented by Kuwahara et al. (2005), who proposed a microscopic traffic model in which certain environment parameters, like traffic accidents or congestions caused a human reaction into driving behaviors like lane changing, speed changing, route choice, etc. This work concludes that such local changes affect macroscopic behavior of traffic. A very important way to simulate traffic simulations is to combine geographic information systems (GIS) and virtual-reality models in order to give more detailed information, as described by Raghothama and Meijer (2015). They proposed a simulation based on SUMO traffic simulator combined with a pedestrian simulator in order to obtain a more accurate environment. Their model allowed simulation of different urban scenarios and pedestrian behavior.

An important use of traffic simulations is in the transportation systems, in order to predict the effects of certain policies to facilitate carpooling or car sharing, for example. In this way, Beutel et al. (2015) presented an agent-based model in which behavior of people was influenced by incentives focusing on sharing transport or bicycle use. Their proposed model can be used to evaluate the impact of incentives, new regulatory rules in areas as ambient impact, traffic density, economy, and others.

2.1 Road Traffic Simulation Platforms

There are several traffic simulation platforms, among which those that are described below stand out:

- *VISSIM*: This traffic simulator is commercial software, widely extended with more than 7,000 licenses distributed worldwide (Barceló, 2010). VISSIM is a software focusing on microscopic traffic simulations and allows modeling of highly complex road scenarios including different agents, such as pedestrians, traffic lights, various types of cars as well as various means of public transport. These allow evaluation of the possible impact before new strategies or mechanisms are implemented on the roads. In this way, it is a powerful tool in decision making on roads and transportation in large cities (PTV Group, n.d.).

- *AVENUE*: *Advanced & Visual Evaluator for road Networks in Urban arEas*: AVENUE is a software that originated in Japan in the 90s. It is based on a hybrid model, that is, the flow model is dynamic, but the visualization of cars is discrete for a better understanding. It is applied for the modeling of small-to-medium scenarios (up to 100 intersections) (Barceló, 2010). AVENUE focuses on the simulation and measurement of the impact of new urban developments as well as strategies focusing on vehicular traffic, such as advanced control of traffic signal timing and synchronization. It has a graphical interface that allows the user to add elements, such as background images, road construction, traffic-flow configuration, among other features. It allows to import and export of data in certain formats and its object-oriented scheme allows a detailed configuration (i-Transport Lab. Co., Ltd., 2018). AVENUE has a tendency to be used as a support tool for research.

- *Paramics*: This has its origins in a transport consulting company in Edinburgh, Scotland. The first commercial version of this simulator dates back to the year 2000. It focuses mainly on the modeling of transport projects, which allow working from a level of urban micro-scenarios to a macro level where regional logistic schemes are evaluated. Its strength lies in the modeling of micro scenarios that allow the configuration of incidents and analyze their impact. Its construction philosophy is that transport must reach a certain destination in any possible way (Barceló, 2010). Currently, it allows simulations in 3D environments and configuring of a wide range of parameters, such as traffic lights, variable traffic density, construction of new infrastructure and its possible impact, eventualities in public transport and the economic and environmental impact (emissions) of these parameters. The data thrown by their simulations are easy to interpret, thus facilitating their understanding by anyone. It takes care of a very user-friendly interface so that its use becomes very intuitive, allowing the user to gain experience in its use quickly (SYSTRA, n.d.).

- *AIMSUN*: *Acronym of Advanced Interactive Microscopic Simulator for Urban and Nonurban networks*: Created at the University of Catalonia (UPC), it was originally conceived as a focused software but at present, it is a software of commercial use, belonging to the Transport Simulation Systems (TSS) company. AIMSUN currently supports simulations in micro scenarios, medium and macro scenarios, allowing modeling and simulation from a car to half of a city. It is one of the most successful commercial simulators, allowing offline analysis of traffic engineering as well as real-time decisions and measures their impact. It allows evaluation of parameters, such as the route of lesser or greater weight (time, costs, tolls) under a modularity scheme, which allows exchange of information between micro-, medium- and macro-levels, thus detecting faults or duplicates of information in a very efficient way. It allows configurement of environmental characteristics, such as railway crossings, lanes dedicated to cycle ways or reserved, pedestrian crossings, among others (Siemens, 2017). Currently there are two versions. AIMSUN focuses on simulations of urban scenarios of different degrees of complexity as well as the effects of strategies, such as shared cars, risk analysis, economic impact of toll booths, environmental impact, confined lanes among other mechanisms. The other version, AIMSUN allows the simulation of medium-size scenarios, simulate high traffic density situations and evaluate the impact of strategies to follow.

- *MITSIMLab*: *Microscopic Traffic SIMulation Laboratory* is a microscopic traffic simulator designed to evaluate the impact of alternative traffic management system designs and assist in their refinement (Barceló, 2010). Examples of scenarios that can be simulated are public transport operations, route guidance systems and others. MITSIMLab compiles and runs under Linux operative systems and is an open-source software, developed at the MIT Intelligent Transportation Systems (ITS) Program. MITSIMLab is supported by a computer community in a forum; any change in original packages can be posted in user's group website. This traffic simulator system contains three main modules (Intelligent Transportation Systems Laboratory, n.d.):

  ○ *Microscopic Traffic Simulator (MITSIM)*: It is in charge of representing the world, in means, traffic, and network elements as represented here in order to capture the sensitivity of traffic flows to the control and routing strategies.

  ○ *Traffic Management Simulator (TMS)*: This is responsible for simulating the traffic control system to be evaluated.

  ○ *Graphical User Interface (GUI)*: This module allows the user to interact with the system, showing traffic impacts through vehicle animation.

- *SUMO: Simulation of Urban Mobility*: This is a microscopic road-traffic simulation available under the GNU General Public License, both as a source code and in compiled executable form for Windows and Linux Systems since 2001. SUMO allows modeling of inter-modal traffic systems, including road vehicles, public transport, and pedestrians. SUMO supports tasks, such as route finding, visualization, and network import and emission calculation. SUMO can be enhanced with custom models and provides various APIs to remotely control the simulation. To use SUMO for research purposes, users are requested to include DLR German Aerospace Center in the acknowledgement. The SUMO platform offers many features, such as microscopic simulations, different traffic types, online interaction, traffic lights timers and others. It is implemented in C++ and uses portable libraries (DLR German Aerospace Center, n.d.). Other features of SUMO platform include small memory footprint; large scenarios must be simulated in a single workstation and the simulation required time is as small as possible.

- *DRACULA*: *Dynamic Route Assignment Combining User Learning and microsimulation*: This traffic simulation model has been developed at the University of Leeds since 1993. DRACULA was developed as a tool to investigate the dynamics between demand and supply interactions in road networks. The DRACULA model attempts to represent directly the behavior of individual drivers in real time as this changes and evolves from day to day. It uses a detailed within-day traffic simulation of the continuous movements of individual vehicles according to car-following and lane-changing rules and traffic controls. The full DRACULA framework is composed of several sub-modules of traffic flow and driver's behavior. The system evolves continuously from one day to the next until a pre-defined number of days, or a broadly balanced state between the demand and the supply, is reached. The essential property of the DRACULA traffic simulation model is that the vehicles move in real-time and their space-time trajectories are determined by car-following and lane-changing models for example. However, vehicle simulation in DRACULA interacts strongly with, and is influenced by, the requirements of its demand model. For example, the explicit modelling of individual drivers' day-to-day learning has an impact on the length of the simulation time-period (Barcelo, 2010; Institute for Transport Studies, n.d.).

- *TransModeler*: This is a traffic simulation system capable of simulating a wide range of road scenarios, such as highways, road cruises, bridges, among others. It allows interaction between different agents, such as pedestrians, vehicles of different types, intelligent transport systems, traffic lights to configure models of high complexity and diversity. TransModeler allows integration of geographic information systems (GIS), thus facilitating the construction of models very attached to reality, both in appearance and behavior. It allows microscopic, mesoscopic, and macroscopic simulations through a hybrid scheme, interconnecting diverse micro scenarios to generate a macro scenario. The results of the simulations allow us to analyze a large number of variables of the analyzed system, such as the degree of use of a highway, vehicular density on an avenue, average cruise travel speeds, average travel times, among many others (Caliper Corporation, n.d.). Licenses vary according to maximum number of intersections allowed to simulate into scenarios.

It should be remembered that in this chapter, the AnyLogic Platform is used because it offers different elements in 2D and 3D, as resources for modeling and simulation, as well as basic libraries of functions and packages of classes and interfaces in Java. For the programming of behavior of the agents, AnyLogic Platform allows the possibility to specialize and extend the functionality of the agents when importing own authorship libraries developed by the programmer.

3. Analysis of Control and Reduction Vehicular Traffic System

The main cause of traffic congestion is that the volume of traffic is too close to the maximum capacity of a road. For example, the congestion in Mexican cities is worse than many, perhaps most, of the other Latin American cities. In addition, the most worrying thing is that this problem is getting worse, year after year.

Traffic congestion occurs mainly because of the increase in the number of vehicles competing for the limited space available on roads. Drivers traveling from any origin to any destination tend to choose the roads that are on the route with the shortest travel time. This behavior results in the underutilization of some roads, and the congestion of others having more vehicular traffic than the one these can contain, or for which these were designed, resulting in imminent traffic congestion. This is because vehicular traffic is growing faster than road capacity. In the absence of measures to reduce vehicular traffic, the problem will continue, since it is impossible to match a road program that addresses the growth of vehicular traffic without restrictions. Under current social and economic frameworks, there are no feasible policies that can reduce the traffic congestion to zero.

In essence, it is necessary to develop a virtual model that represents an appropriate control system for road traffic, in order to study the causes and effects of vehicular congestion or traffic jams for its control and reduction, based on good and effective alternatives or strategies proved virtually and offering the best impact in real life.

3.1 Strategies to Control and Reduce Road Traffic

Currently and for decades, in large urban areas during periods of increased demand, known as peak hours, vehicular congestion is inevitable. Although different strategies have been implemented for traffic control and reduction, the costs these impose are usually high. Many strategies have failed due to the used methodology, since the programs carried out lack a previous analysis that provides a correct orientation to be successful in practice.

Thomson (2000), within the project 'Charging for Use of Road Space in Latin American Cities', exposed some of the most relevant causes of traffic congestion, saying the following:

- *It is the cars that have created the greatest problem*: A car has less capacity to transport passengers and most of the time they only carry one, and that is the driver. This is contrary to public transport, which can transport a greater number of people and their routes are made in more time.

- *Some driver behaviors cause more vehicular traffic*: In Latin American cities, behaviors are observed in drivers who park on the avenues and before road intersections, preventing free transit of other vehicles.

In addition, within the mentioned project, the different strategies that have been applied to counteract vehicular traffic were discussed and which are also addressed by Bull (2003), highlight the following:

- *To grow and provide more road infrastructure*: This has not worked since the problem of vehicular congestion during peak hours persists.

- *Expansion of the public transport network*: In cities that have subway lines, the desire to transport a greater number of passengers collectively and en masse, has not significantly reduced traffic congestion, since users use an additional means that takes them from other places to subway stations or vice versa, generating a greater vehicular traffic in the surrounding environment.

- *To improve public transport and encourage less use of the car*: There have been doubts as to whether this strategy is attractive for motorists, since it is more comfortable to travel by car than to do it in an uncomfortable public transport.

- *To have a road fare*: This implies having to impose a fee on those wanting to reduce their transfer times on private roads with greater traffic flow. However, these are not enough and have presented the similar problem of over-saturation during peak hours.

- *Staggered hours*: This is very difficult to evaluate and to put into practice, since it supposes that the right of free movement for motorists is subtracted or removed at certain times of the day in areas of great traffic congestion.

- *Actions on road crossings*: A better design of intersections with adequate control systems, to have adequate returns and roundabouts for a better distribution of vehicular flow are projected. But this usually causes congestion on road crossings for both those entering and those leaving thereof.

- *Lanes dedicated to public transport*: This offers a reduction in transportation time for public transport but leads to a reduction in lanes and decreases the capacity of transportation for motorists. During the peak hours, vehicular traffic worsens instead of improving.

3.1.1 Government Programs

In countries like Mexico, there is the 'today does not circulate' program, in which certain vehicles do not circulate one day of the week. However, it has not worked properly either, because many motorists prefer to buy an additional car to be able to drive throughout the week. In addition, programs for the use of alternative means of transport, such as bicycles, have been promoted since the last decade, but these programs require investment in new infrastructure works to enable cycle ways on roads, causing vehicular traffic problems during its construction and a reduction in road capacity. This results in increased vehicular traffic.

What appears to offer a better way for more promising results is to create strategic programs under a methodological approach of a package of measures that look for the following:

- To take advantage of the existing infrastructure and before its growth, see if new investment costs are necessary.
- To efficiently transfer time through the synchronization of traffic lights to reduce traffic congestion rates. This demands intelligent signaling and at the same time be auto adaptive, to the needs of control and reduction of vehicular traffic 24 hours a day and seven days a week, without interruption.
- To understand what has been described, simulation offers a powerful tool that, when handled in an appropriate manner, would give the desired answers to large premises, such as knowing whether a strategy would work or not. Here it is possible to take advantage of a better analysis and design for an adequate system of control and reduction in vehicular traffic, which can be improved under the advantages of modeling and simulation.

3.1.2 Module Specification

To improve traffic congestion in a successful way, the simulation previously requires a correct division of modules in the analysis phase for a good definition in the modeling process.

There are three variables that define the characteristics of vehicular traffic: flow, speed and density, which are used for the operation of existing transit systems. However, to optimize the control systems and reduce vehicular traffic, it is necessary to have at least the following specifications of modules for a correct modeling and a good simulation (Fernández, 2008):

- *Road management*: Controlling the capacity of roads and their degree of saturation.
- *Optimization of road intersections*: Synchronization of traffic lights, use of priority signs, and distribution of vehicular traffic through roundabouts are necessary.
- *Passenger transfer*: Public transport stops and subway stations.
- *Pedestrian enablement*: Pedestrian crossings at traffic lights or established steps, such as zebra crossings.
- *Support to motorists*: The use of information monitors information about the current traffic situation in areas of greater congestion so that the use of alternative routes is anticipated.

3.2 Creation of a Vehicular Traffic Model Using the AnyLogic Platform

The vehicular flow occurs at different times and places, depending on road conditions and the decisions made by drivers. Each driver at the wheel takes the route that considers the best or the shortest to reach a destination. These decisions can be made depending on the time of day, the distance from one place to another, the safety on the journey, or the season of the year. There are times when vehicular traffic is dense due to hours of entry to schools and jobs or it may be light because it is night time or the early hours of the morning. Users employ various means of transport, such as cars, motorcycles, bicycles, public transport, among which some may be faster than others, depending upon the quality and speed of road traffic or current weather conditions.

The simulation by a virtual vehicular traffic model allows a simpler treatment of the global behavior, based on individual decisions and conditions that may arise during a journey and in this way, to know the characteristics that a road or street must have. Using the AnyLogic Platform, it is possible to represent graphically the actors and factors involved in a vehicular traffic system under 2D and 3D modeling. The AnyLogic Platform University 8.2.3 version was used here.

3.2.1 AnyLogic Platform Road Traffic Library

The 'AnyLogic Platform Road Traffic Library' allows to obtain a detailed physical level when planning, designing, and simulating vehicle traffic flows and managing road traffic systems, as a transport-planning tool and vehicle traffic engineering. Road traffic models simulate traffic on streets and highways, including crossings or intersections, pedestrian crossings, roundabouts, parking lots, and bus or public transport stops. As presumed, AnyLogic Road Traffic Library is ideal for the following:

- Explicit modeling of the behavior of each vehicle or driver to represent the dynamics of vehicular flow in road networks.
- To respect typical driving rules, such as speed control, the choice of the least busy lane, the rules for lane combination, as well as avoiding and detecting collisions.
- To represent each vehicle as an agent that can have its own physical parameters highlighting the length, speed and acceleration, as well as their behavior patterns that are simulated with flow diagrams that can be easily constructed. Capturing behavior is crucial when evaluating the performance of the entire traffic system.
- To create 2D and 3D models of each vehicle and its environment so that traffic models are 3D flexible for better visualization.
- To easily model intersections or intersections under priorities, traffic lights, pedestrian crossings, bus stops or public transport stops and parking lots.
- To have traffic density map to visualize traffic congestions and collect statistics on traffic flows.
- To convert files from geographic information systems, with data on existing roads, to AnyLogic Platform road-space marking forms. In this way, the road network is drawn automatically.
- To expand the vehicular traffic model with the elements of the AnyLogic Platform Pedestrian Library (AnyLogic, Pedestrian Library, n.d.) and the AnyLogic Platform

Rail Library (AnyLogic, Rail Library, n.d.) to simulate complex transportation systems, including railway centers and airport terminals.

The capabilities of this library are instrumental when the following is done:

- Road planning.
- Evaluation of the capacity and performance of the road network.
- Management of the level of traffic congestion.
- Completion of the semaphore schedules.
- Integration of public buildings in a road network.
- Support to various types of cars to customize their animation and attributes.

It is worth mentioning that at the time and until the date in which this section was written, the library does not support the reverse movement of cars and lanes so that cars can move in both directions. However, based on the help manual given in (AnyLogic, Help AnyLogic, n.d.), it contains the following seven blocks which allow defining the vehicular traffic flow into a virtual simulation model:

1. '*CarSource*': It generates cars and places them in a specific location within a road network. Car arrivals can be defined by inter-arrival time, arrival rate, rate schedule, arrival time or calls to the inject function. In addition, the number of arrivals can be limited. The following considerations have to be taken into account:

   a. If cars are configured to appear on a road, the direction of the road must be specified additionally. Cars will appear at the beginning of the specified route and start to move in a specified direction. By default, the random rail parameter is selected, so that cars will appear in random lanes (if there is more than one lane in the specified road). The lane index, where cars will appear by de-selecting the random lane option and using the lane index parameter, must be explicitly specified.

   b. If cars are configured to appear on a road, they enter the road network only when there is enough distance in the lane in front of the car. The distance needed for a car to enter the network safely depends on the initial speed of the car. Cars that cannot enter the road network immediately are queued within the 'CarSource' block and are removed from the queue when there is enough space in the specified lane to place a car.

   c. If cars are configured to appear in a parking lot, they appear in free spaces of the specified parking lot and wait until they leave the road network, or begin to move when entering the 'CarMoveTo' block. If all the parking spaces are occupied, an error will be generated.

2. '*CarDispose*': It removes a car from the model. A car can be removed as follows:

   a. The car can move through any open road or when it reaches the specified stop line, in this case 'CarDispose' it must follow the last 'CarMoveTo' block.

   b. The car can be removed from a model when it is located in a parking lot or at a bus stop. In this case, it is not necessary to remove the car immediately after it arrives at the place.

3. '*CarMoveTo*': It controls the movement of the car. A car can only move while it is inside this block. When a car enters the block, it calculates the road from its current location to the specified destination. The destination can be a road, parking area, bus

stop or stop line. If the destination is a road, the car will first travel along the shortest path to the beginning of the road and then proceed along this road to the exit. If there is no road from the current location of the car to the specified destination, the car leaves the block through the 'OnWayNotFound' port. Next, some possible motions for a car are described:

a. If a car is configured to move to a road, the direction of the road must be additionally specified. The car will try to use the shortest route to get to the beginning of the specified road. Once in it, the car will move along the specified route to the end. Once a car reaches the end of the road, it must be immediately removed from the model by means of the 'CarDispose' block or by going to the other 'CarMoveTo' block to send it more. If an automobile cannot leave the end of the destination lane immediately, an error will be generated.

b. If a car is ready to move to a parking lot, the car will move to the specified parking lot and will attempt to reserve a free parking space. Cars reserve parking spaces only when they approach the parking lot. If a car cannot reserve a parking space, it leaves the block through the 'OutWayNotFound' port. If a car can successfully reserve a parking space, it parks and leaves this block. A car can remain in the parking lot until it is removed from the road network or sent to another destination by the 'CarMoveTo' block.

c. If a car is configured to move to a bus stop, the car will move to the specified bus stop. If a bus stop is already occupied by another car, the car approaching the bus stop will stop at the bus stop and wait until the bus stop is vacant. When a car arrives at a bus stop, it parks and leaves this block. A car can remain at the bus stop until it is removed from the road network, or sent to another destination by the 'CarMoveTo' block.

d. If a car is configured to move to a stop line, two behavioral options are possible on the destination stop line (see the stop line behavior parameter for the description).

e. While the car moves under the control of the 'CarMoveTo' block, it can still be partially controlled using the 'Car API'.

4. '*CarEnter*': It takes the automobile agent and tries to place it like a car in the specified location within a road network. The car may appear in the specified way or in the specified parking (defined by the parameter Appears). Together with 'CarExit', this block is used to model a part of the motion of the car at a higher level of extraction, that is, without a road traffic model of detailed physical level.

5. '*CarExit*': It removes the car from the road network and passes the automotive agent to the flow diagram of the AnyLogic Platform Process Modeling Library, where it can go through delays, queues, decisions, etc. Together with 'CarEnter', this block is used to model part of the movement of the car at a higher level of abstraction; that is, without a detailed physical road traffic model.

6. '*Traffic Light*': It is used to simulate one or more semaphores. A traffic light, also known as traffic signal, signal light, braking light, is a signaling device located at road intersections or crossings, pedestrian crossings and other locations to control conflicting and heavy traffic flows. A 'traffic light' block controls vehicular traffic at a specified intersection or stop line.

7. '*RoadNetworkDescriptor*': This is an optional block that gives access to the control of all vehicles located in a road network. It allows to configure the actions that will be executed by each car in the following cases—when entering the network, when entering the road, when changing lanes, etc. This block also allows seeing a road density map that shows the current state of traffic jams on the road network.

3.2.2 Color Animation of the Flowchart

The AnyLogic Platform animates all the flow diagrams at execution time of the model. In Fig. 1, it can be seen that the color scheme for most of the blocks in the model window includes three colors: silver, dark gray and gray, to distinguish the current state of the block.

By default, all blocks are colored under a basic scheme in the following way:

- *Silver*: It indicates that cars have not yet entered the block.
- *Dark gray*: It indicates that cars have entered this block earlier, but it is currently empty; that is, it does not contain cars inside.

The processing of the actions of the automobiles that takes place inside the block is visualized as follows:

- *Dark gray with gray*: It indicates that the block currently contains cars inside and an operation is being processed.

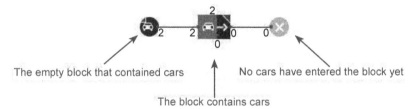

The empty block that contained cars   No cars have entered the block yet

The block contains cars

Fig. 1: Color animation code to distinguish the current state of a block (*Source*: AnyLogic, Road Traffic Library, n.d.).

3.3 Modeling a Real Road Crossing in AnyLogic Platform

This section will show how to model the motion of automobiles on the road network of a vehicular cruise, known as 'Puerto de Chivos', which is located between the boundaries of the municipalities of Atizapan de Zaragoza and Nicolas Romero in Mexico state, Mexico. This cruise is very close to the University Center UAEM Valley of Mexico of the Autonomous University of Mexico state. This University Center has the Artificial Intelligence and the High Performance Computation Laboratories where the authors of this chapter work and do research activities as the one presented in this chapter.

Here, a virtual vehicular traffic model will be created over an image obtained and adapted from the screenshot given in a satellite way, by Google Maps for the aforementioned road junction. This image is shown in Fig. 2, where all the peculiarities of the road network of the crossing are appreciated. The roads are mostly bidirectional and contain one or more lanes for each direction of motion. In addition, there are traffic lights and bus stops or public transport. Systematically, all these peculiarities will be considered

Fig. 2: Satellite image of the Puerto de Chivos road junction in Atizapan de Zaragoza, Mexico state, Mexico.

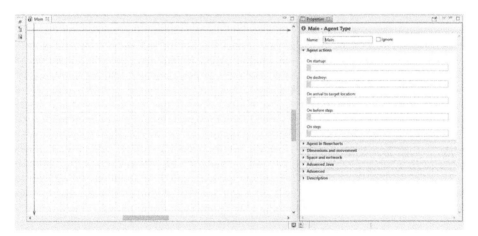

Fig. 3: Editing view to design and configure a created model in AnyLogic Platform.

in the vehicular traffic model to be created, demonstrating the use and application of most of the blocks or elements of the AnyLogic Platform Road Traffic Library, as it progresses in its creation and implementation.

For the creation of a virtual vehicular traffic model in AnyLogic Platform for the mentioned road crossing, the following steps were carried out:

1. A design was added and the graphic editor was configured in a new created model, where elements and blocks used as components were added mainly from the AnyLogic Platform Road Traffic Palette. Figure 3 shows the graphic editor of the created model; agents are the main building blocks of a model in AnyLogic

Fig. 4: Satellite image added as design to the canvas into the editing view of the created model.

Fig. 5: AnyLogic Platform Road Traffic Library Palette.

Platform. In this case, the 'main' agent will serve as the place where the logic of the vehicular traffic model is defined, since the road network is drawn on it and the vehicular traffic process is defined through a flow chart. In Fig. 4, the satellite image of the Puerto de Chivos road junction was added as a design in the graphic editor.

2. The 'AnyLogic Platform Road Traffic Library Palette' was used to draw and configure a road network, placing exactly each road on the design over the corresponding road in the satellite image, thus respecting the flow direction of vehicular traffic

Fig. 6: Roads drawn to create the road network.

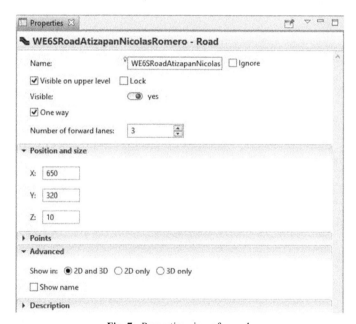

Fig. 7: Properties view of a road.

(Figs. 5 and 6). Figure 7 shows the way to configure each road for one way or two ways, as well for its number of lanes in each direction.

3. Crossings or intersections, created to connect roads in AnyLogic Platform, were drawn and created automatically when drawing them one by one, which creates a connection in order to join two or more roads. Figure 8 shows the crossing or intersection that is automatically generated on the road network and its properties as well. To configure the direction of traffic flows from one road to another through an intersection, AnyLogic Platform allows establishing of the possible lanes of union and continuity as flows of circulation. These flows can be activated and deactivated

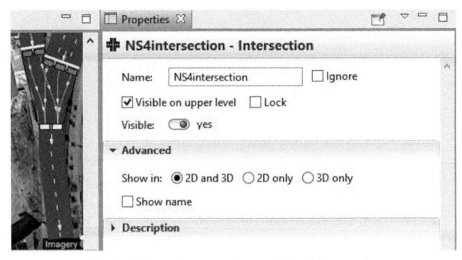

Fig. 8: An intersection generated automatically to join two roads.

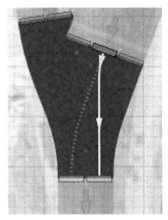

Fig. 9: Activation/deactivation of traffic flows between roads on a cruise or intersection of roads.

by clicking on them. With it, a flow with continuous line indicates that it is activated and one with dashed line indicates that it is deactivated (Fig. 9).

4. Parking lots were added to perfect the model of the road crossing. This made possible to model cars that occasionally run and stop for a while in a parking space. The steps of the process of creating and drawing parking lots are the following: to drag the 'Parking' element of the AnyLogic Platform Road Traffic Library Palette to the road where it will be placed. A parking lot can be drawn only if the road has been previously drawn. Figure 10 shows that by hovering the mouse over the road network, all the other elements, except the roads and its direction flows, are temporarily hidden to appreciate where exactly the parking lots will be placed. Parking lots are automatically connected to roads and will adjust their shape to the form of the road. A parking lot is configured as a destination in the 'CarMoveTo'

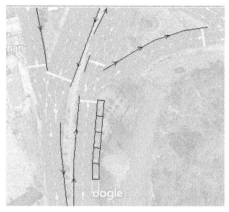

Fig. 10: Dragging and placing parking lots on a road.

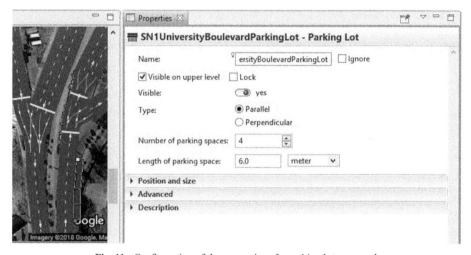

Fig. 11: Configuration of the properties of a parking lot on a road.

block for simulating the motion of a car to enter a parking lot, stay for a specific time, and leave the lot to continue its direction. Figure 11 shows the way to modify properties, such as the number of parking spaces for a parking lot.

5. Bus stops were added in the direction of displacement of the vehicle flow to simulate public transport stops. The 'CarMoveTo' block was used to model the motion of buses to particular desired stop places and simulate stops for a certain period. To complement this, it was needed to use the 'Delay' block of the AnyLogic Platform Process Modeling Library. Figure 12 shows that hovering the mouse over the road network, all the elements, except roads, are temporarily hidden to appreciate where exactly the bus stops will be placed. In a similar way to parking lots, bus stops are automatically connected to roads and will adjust their shape to the form of the road. In addition, bus stops are configured as destinations in the 'CarMoveTo' block for simulating the movement of public transport to enter a bus stop, stay for a specific

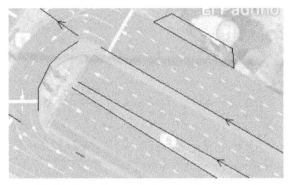

Fig. 12: Dragging and placing a bus stop on a road.

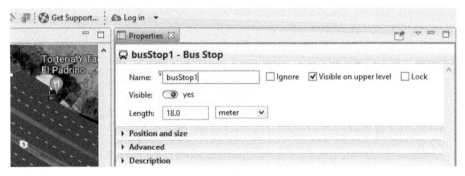

Fig. 13: Configuration of the bus stop properties.

Fig. 14: A full road network on the satellite image of the road crossing is depicted.

time and leave it to continue its direction. Figure 13 shows the way to modify properties, such as length of a bus stop.

6. The road network was completed by adding all roads, intersections or crossings, parking lots and bus stops, as well as activating and deactivating circulation flows and configuring their respective properties as displayed in Fig. 14. The road network

were narrower or wider than the corresponding one on the satellite image, with the scale of the model adjusted in its properties. This was done by adjusting the scale to set 25 meters for each 100 pixels (Fig. 15).

3.3.1 Creation of Agents to Represent Vehicles

Inside the virtual created model of vehicular traffic, only particular cars or automobiles and public transport were considered as agents, leaving aside others, such as motorcycles and bicycles that also circulate on the real road network. Agents were created from scratch and customized by choosing the type of 3D animation through selection of different types of road transport as shown in Fig. 16.

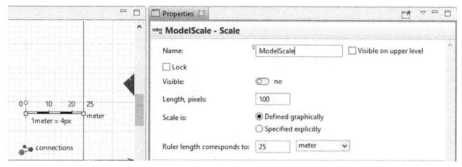

Fig. 15: Scale element adjustment in the created model.

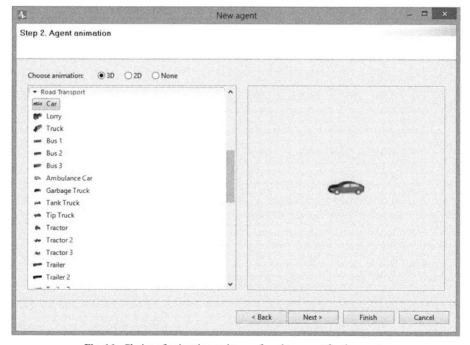

Fig. 16: Choice of animation and type of road transport for the agents.

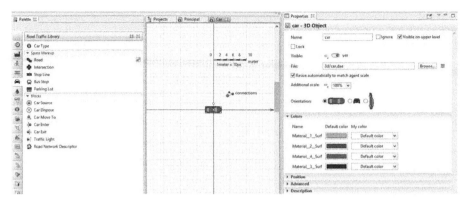

Fig. 17: Adjustment and configuration of properties of the 3D representation of the agents.

In order to have several presentations of the types of road transport in different colors and adjust their size for the agents, a random way to animate was established in the model at runtime, which means that they will appear in different types, sizes and colors randomly as seen in Fig. 17.

3.3.2 *Vehicle Flow Diagram for the Virtual Road Network Model*

The vehicular flow for the virtual vehicular traffic model was modeled and created with a flow diagram composed of blocks of the AnyLogic Platform Road Traffic Library Palette. Blocks were used to define operations or actions for the agents represented as vehicles that circulate through the virtual road network. In AnyLogic Platform, vehicular flow diagrams are created by adding and connecting as many blocks as required of each type of the AnyLogic Platform Road Traffic Library Palette in the graphic design of the vehicular traffic model that is being built, and by adjusting its parameters.

Additional details of the three most used blocks are:

- '*CarSource*' *generates cars*: In general, it is used as a starting point of the flow chart to indicate where a car appears and starts to circulate on the road network generated.
- '*CarMoveTo*' simulates and controls the movement of a car to a specified destination.
- '*CarDispose*' indicates the end of the flow diagram and discards the cars that arrive from the model.

The configuration of properties for a block of type 'CarSource' can be seen in Fig. 18. Figure 19 shows the flow diagram that was created to give the road network a flow configuration in the vehicular traffic model generated to represent the Puerto de Chivos road junction. This configuration indicates a flow logic that defines the movement of vehicles through roads and intersections on the virtual road network. With the help of the 'SelectOutput' or 'SelectOutput5' blocks of the AnyLogic Platform Process Modeling Library, different routes and addresses were assigned to the vehicles.

3.3.3 *Placement of Traffic Lights*

In order to control the passage of vehicles from one road to another at the intersections, traffic lights were added to the vehicle-flow diagram. The same also applies to pedestrian

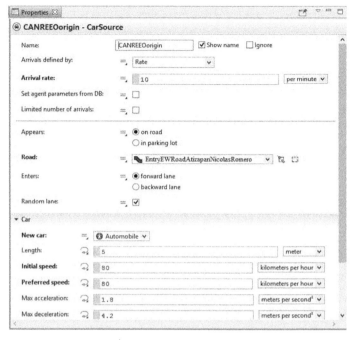

Fig. 18: 'CarSource' block and its properties.

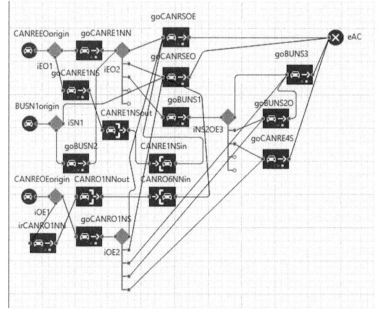

Fig. 19: Vehicle flow diagram for the virtual road network.

crossings, but this was not done here. The 'Traffic Light' block of the AnyLogic Road Traffic Library was used to define the logic for the traffic lights (Fig. 20):

Attention must be paid to the main block parameter 'Defines the mode for'. This parameter has three alternative options described as follows:

1. The 'Intersection's stop lines' option, selected by default, is the easiest to configure, since the traffic lights created will control traffic by enabling/disabling traffic flow in all lanes of the road or road adjacent to the intersection. This option was used in the traffic lights of the vehicular traffic model generated.

2. The option 'Intersection's lane connectors' is used when the allowed traffic directions at the controlled intersection differ for lanes of the same road during the same traffic light phase.

3. The option 'Specified stop lines' is used when the traffic lights are simulated by this block to control the traffic on pedestrian crossings:

    1. In the 'intersection' parameter, choose the name of the intersection to be controlled.

    2. Set the duration time in seconds of the red and green phases for the stop lines and track connectors.

    3. If required, the phases that are necessary can be added.

When selecting a cell from the 'Phases' table of the traffic light, all the shapes in the graphic editor are dimmed, while stop lines or track connectors are marked in red. This means the editing mode of the phase is activated, allowing easy change in the status of each stop line for the current phase directly in the graphical editor (Fig. 20). The color will change from red to green and the corresponding cell in the 'Phases' table will also change to the same color.

The idea of using phases, their colors, and their duration is in order to switch traffic light phases, which regulate the traffic flow at intersections.

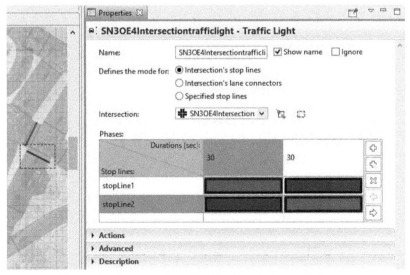

Fig. 20: Logic change of a traffic light phase in the graphical editor.

3.3.4 Configuration of the Direction of Vehicular Flow in the Created Virtual Road Network

A road network is built through an in-depth search of the connected road segments (lines and arcs). Two lanes are considered connected if their ends are closer than the connection tolerance. A road direction is determined by the orientation of the corresponding form. To assign the correct direction to the vehicular flow in a road network, it must be established whether it is on the right or left side. In the case of the vehicle traffic model created here, this direction was on the right side. In countries like Singapore, this sense must be on the left side.

The number of lanes in a road is calculated as the width of the road in meters divided by the width of the lane. Therefore, it is necessary to provide the scale parameters that convert pixels into meters. The color of the surface of the road network and the marking can be customized under the parameters of 'Appearance'.

Recalling Fig. 14, the configuration given for the road network of the generated model was a value of 3.5 meters established for 'Lane width' and for 'Road color', the texture 'tarmac' that gives the effect of a paved road network. The rest of the properties and parameters were left with their default value.

If all the elements and objects placed as a group of shapes are 3D, the 'Road network' element will create an animation in three dimensions, otherwise a 2D animation is shown. When choosing the 3D display option of a group of shapes, the height on the 'Z' axis is automatically adjusted to 10, which may not be the desired height of the roads. The easiest way to change this height is in the content. For this, it is necessary to establish in all roads the value of the 'Z' axis at 10, or at an appropriate size (Fig. 21).

The object 'Road network' does not automatically detect road junctions; that is, places where two roads cross each other and without connection. So, if the crossing is of a single plant, it is necessary to make good decisions to ensure that there are no vehicle collisions at execution time.

Fig. 21: Height value on the 'Z' axis for roadways.

3.3.5 Descriptor of the Virtual Road Network

Once the road flow diagram was completed, the number of times the lights of each phase in the traffic lights was configured, the direction of the vehicular flow was established. The 'RoadNetworkDescriptor' block can be placed optionally on the design so that the following can be done (Fig. 22):

- To set the property 'Road section length' of the selected road network.
- To work with density map parameters.
- To define the decision making on different actions in the road network at execution time of the model.
- To operate with additional actions to resolve conflicts and maintain accuracy over calculations.
- To set parameters of the type of agent that will be used to represent vehicles on the road network.
- To give a description of the block.

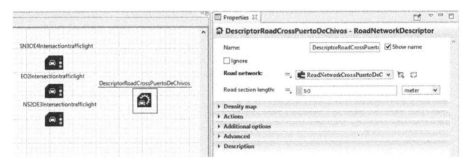

Fig. 22: Descriptor of the virtual road network.

3.3.6 Traffic Density Map

AnyLogic Platform allows to show a colored map of vehicle density or vehicular traffic congestion to identify regions liable to suffer from traffic jams in the simulated road network, at the execution time. Through the map, the roads are painted in different colors gradually as vehicles move on the roads. The color of each road segment corresponds to the current car density on it. The density map uses colors from green to red to show different car densities. The red color is used to show segments of roadways with critical densities that have a congestion problem or traffic jam. The green color is used to show road segments with normal traffic.

A density map is constantly repainted according to the present situation. When the density changes in some segment of the road, its color changes dynamically to reflect this change. By default, the vehicle density map is not shown, but can be activated with just one click on the properties of the density map section of the object associated with the 'RoadNetworkDescriptor' block of the virtual model.

In Fig. 23, the way to configure and customize the density map is shown, setting the different speed levels for traffic lights, and the transparency as well for a specific road section.

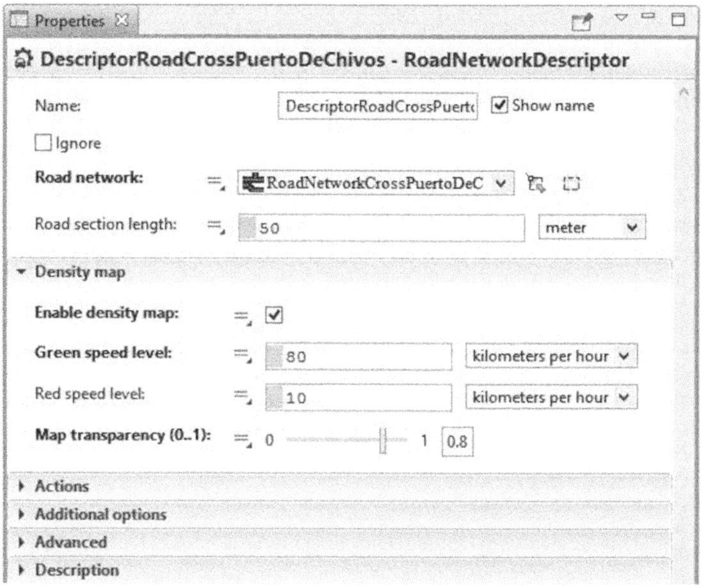

Fig. 23: Customization of the vehicle density map of the virtual road network.

3.3.7 Enabling a 3D Model Animation

In order to create a 3D animation of the vehicular traffic model, first, a 3D window was added. A 3D window plays the role of a placeholder for 3D animation. This window defines the area in the presentation diagram where 3D animation will be displayed at run time. In Fig. 24, a 3D view of the model in execution is given and thus the vehicular density map can be appreciated.

Fig. 24: 3D execution of the vehicular traffic model and its vehicular density map.

3.3.8 Evolution of the Virtual Road Network to Represent a More Realistic Scenario

So far, what has been developed does not show a realistic 3D virtual scenario; it only allows appreciation of a view of the superposition of the 3D representation of the agents that represent vehicles on the 2D image of the modeling road crossing. To achieve a better 3D scenario respecting as much as possible what exists, converges, and diverges in the environment and surroundings of the Puerto de Chivos crossing, the virtual road network was evolved to have a better model. Figure 25 shows the execution of the evolved scenario, proving that it was possible to have a complete and complex model to represent in a virtual way the vehicular crossing of Puerto de Chivos. This 3D view allows appreciating a real scale for the elements existing in the environment and the adjacent surroundings to this road crossing.

Fig. 25: 3D view of the execution of the evolution of the scenario in the model.

4. Conclusions and Future Work

Currently, multi-agent systems are an area of growing interest within Artificial Intelligence due to the ability to adapt in order to resolve complex problems that have not been solved satisfactorily by other techniques. One of the problems in large urban areas, such as the metropolitan area of the Valley of Mexico in Mexico state and Mexico City in Mexico, is the vehicular congestion that occurs on the main crossings at certain times of the day. This fact causes problems, among which the most common ones are: the time lost by a driver or motorist in moving from one point to another, higher rates of pollution, fatigue and even stress.

What was done and shown in this chapter is to study the traffic congestion of road crossings. For this, a virtual model for a real road network, around a vehicle crossing, was created based on a multi-agent system in order to study a real scenario of vehicular traffic. The multi-agent system was represented as a simplified form of reality through simulation. Each one of the elements of the simulation were autonomous agents, which allowed experimenting and simulating real vehicular traffic conditions for the known Puerto de Chivos highway crossing.

Considering that the effects and problems that derive from vehicular traffic are constantly growing, the continuity of this research work is to look for solutions and alternatives, tested in virtual scenarios and which if carried out in real life, would help to minimize those problems. To enrich the simulation and have a better understanding of the vehicular traffic phenomenon and looking for possible courses of action for its control and reduction, what can be done to extend this work is mainly the following future work as follows:

- To create a scenario for a more complex and extensive road crossing.
- To incorporate more types of vehicles.
- To allow seeing and appreciating the part of pedestrians and the pedestrian-driver relationship.
- To use the AnyLogic Platform Process Simulation Library to have a greater freedom in agent behavior.
- To use cameras at floor level and in the manner of drones to have other views of the vehicular flow.
- To collect statistics of congestion rates for analysis.
- To simulate, at different times of the day and weekdays, scenarios that are closer to the reality of the cruise.
- To work with intelligent traffic light control schemes to balance and distribute the traffic flow on roads along the vehicle crossing.
- To use the obtained experience and knowledge to model more complex scenarios to appreciate and prove effective strategies to control and reduce road traffic. The different scenarios can be created to show the phenomenon of vehicle traffic at different scales, such as microscopic and macroscopic views.

References

AnyLogic. (n.d.). AnyLogic Platform, retrieved 02 01, 2019, from https://www.anylogic.com/.
AnyLogic. (n.d.). Help AnyLogic, retrieved 02 01, 2019, from https://help.anylogic.com/index. jsp?topic=%2Fcom.anylogic.help%2Fhtml%2Fmarkup%2F.
AnyLogic. (n.d.). Pedestrian Library, retrieved 02 01, 2019, from https://www.anylogic.com/resources/ libraries/pedestrian-library/.
AnyLogic. (n.d.). Rail Library, retrieved 02 01, 2019, from https://www.anylogic.com/resources/libraries/ rail-library/.
AnyLogic. (n.d.). Road Traffic Library, retrieved 02 01, 2019, from https://www.anylogic.com/resources/ libraries/road-traffic-library/.
Barceló, J. (ed.). (2010). Fundamentals of Traffic Simulation, New York: Springer.
Bellini, P., Bilotta, S., Nesi, P., Paolucci, M. and Soderi, M. (2018). WiP: Traffic flow reconstruction from scattered data. 2018 IEEE International Conference on Smart Computing, pp. 264–266, Taormina, Sicily: IEEE.
Berg, J., Sewall, J., Lin, M. and Manocha, D. (2009). Virtualized traffic: Reconstructing traffic flows from discrete spatio-temporal data. Virtual Reality, pp. 183–190, Lafayette, Louisiana: IEEE.
Beutel, M.C., Addicks, S.S.Z.B., Himmel, S., Krempels, K. and Ziefle, M. (2015). Agent-based transportation demand management. Demand effects of reserved parking space and priority lanes in comparison and combination. 2015 International Conference on Smart Cities and Green ICT Systems (SMARTGREENS), pp. 1–7, Lisbon: IEEE.

Bull, A. (2003). Congestión de tránsito: El problema y cómoenfrentarlo. Santiago de Chile: United Nations Publications.

Caliper Corporation. (n.d.). Caliper, retrieved February 2019, from TransModeler—Traffic Simulation Software, www.caliper.com/TransModeler.

Cavallo, A., Robaldo, A., Bellotti, F., Berta, R., Carmosino, I. and De Gloria, A. (2016). Towards a virtual reality interactive application for truck traffic access management. pp. 169–176. In International Conference on Applications in Electronics Pervading Industry, Environment and Society. Springer, Cham.

Chen, T. and Wei, L. (2011). Virtual simulation test system for traffic safety risks identification. 2011 Fourth International Conference on Intelligent Computation Technology and Automation, pp. 995–998, Washington, DC: IEEE.

Coclite, G.M., Garavello, M. and Piccoli, B. (2005). Traffic flow on a road network. SIAM Journal on Mathematical Analysis, 36(6): 1862–1886.

Diaz, E.M. (2018). Theory of planned behavior and pedestrians' intentions to violate traffic regulations. Transportation Research Part F: Traffic Psychology and Behavior, 5(3): 169–175.

DLR German Aerospace Center. (n.d.). DLR - DLR Portal, retrieved February 2019, from DLR German Aerospace Center: dlr.de/ts/sumo.

Fernández, R. (2008). Elementos de la teoría del tráfico vehicular. Santiago de Chile: Universidad de Los Andes.

Ferrara, A., Sacone, S. and Siri, S. (2018). Freeway Traffic Modeling and Control, Springer.

Haubrich, T., Seele, S., Herpers, R. and Becker, P. (2014). Integration of road network logics into vehicular environments. 2014 IEEE Virtual Reality (VR), pp. 79–80, Minneapolis, MN: IEEE.

Hongke, X., Jinhua, C., Dashan, C. and Na, X. (2007). Simulation of highway tunnel traffic control and guidance. pp. 2-644-2-647. In 2007 8th International Conference on Electronic Measurement and Instruments. IEEE.

Institute for Transport Studies. (n.d.). Faculty of Environment - Institute for Transport Studies, retrieved February 2019, from University of Leeds (GB): https://environment.leeds.ac.uk/transport.

Intelligent Transportation Systems Laboratory. (n.d.). Software – Intelligent Transportation Systems Lab., retrieved February 2019, from ITS.LAB @ MIT. Intelligent Transportation Systems LaB: https://its.mit.edu/software.

i-Transport Lab. Co., Ltd. (2018, December 7). ITL, i-Transport Lab. Co. Ltd., retrieved February 2019, from ITL AVENUE: http://www.i-transportlab.jp/en/products/avenue.html.

Khakimov, Y.M. (2018). Traffic violations: A study of measures aimed at bringing perpetrators to justice. Journal of Advanced Research in Law and Economics, 9(1): 112–118.

Kuwahara, M., Tanaka, S., Furukawa, M., Honda, K., Maruoka, K., Yamamoto, T. and Webster, N. (2005). An enhanced traffic simulation system for interactive traffic environment. Intelligent Vehicles Symposium, pp. 739–742, Las Vegas: IEEE Proceedings.

Mandhare, P.A., Kharat, V. and Patil, C.Y. (2018). Intelligent road traffic control system for traffic congestion: A perspective. International Journal of Computer Sciences and Engineering, 908–915.

Nemec, N., Wlosok, J. and Fasuga, R. (2014). Virtual 3D simulation of a car in traffic. 12th IEEE International Conference of Emerging eLearning Technologies and Applications (ICETA 2014), pp. 349–354, Smokovec, Slovakia: IEEE.

Papageorgiou, M., Diakaki, C., Dinopoulou, V., Kotsialos, A. and Wang, Y. (2003). Review of road traffic control strategies. Proceedings of the IEEE, 91: 2043–2067.

PTV Group. (n.d.). Transportation Planning, Traffic Engineering and Traffic Simulation, retrieved February 2019, from Traffic and Logistic Software & Technology PTV Group: http://vision-traffic.ptvgroup.com/en-us/home/.

Raghothama, J. and Meijer, S. (2015). Distributed, integrated and interactive traffic simulations. pp. 1693–1704. In Proceedings of the 2015 Winter Simulation Conference. IEEE Press.

Siemens. (2017, February 1). Aimsun, retrieved February 2019, from Aimsun, a Siemens Company: http://www.aimsun.com/.

SYSTRA. (n.d.). Paramics Microsimulation—3D traffic simulation brought to you by SYSTRA, retrieved February 2019, from Paramics Microsimulation SYSTRA Ltd.: https://www.paramics.co.uk/en/.

Tang, Y., Zhang, C., Gu, R., Li, P. and Yang, B. (2017, Feb. 1). Vehicle detection and recognition for intelligent traffic surveillance system. Multimedia Tools and Applications, 76(4): 5817–5832.

Thomson, I. (2000). Algunos conceptosbásicossobrelascausas y soluciones del problema de la congestión de tránsito. Santiago de Chile: Naciones Unidas Comisión Económicapara América Latina y el Caribe (CEPAL).

Tian, G. and Akashi, T. (2013). Sensing of traffic flow in real-world for reconstruction in VR. 2013 International Conference on Cyberworlds, pp. 386–386, Yokohama: IEEE.

Wan, T. and Tang, T. (2003). An intelligent vehicle model for 3D visual traffic simulation. 2003 International Conference on Visual Information Engineering VIE, pp. 206–209, Guildford, UK: IEEE.

Yu, Y., El Kamel, A. and Gong, G. (2013a). Multi-agent based architecture for virtual reality intelligent simulation system of vehicles. 2013 10th IEEE International Conference on Networking, Sensing and Control (ICNSC), pp. 597–602, Evry: IEEE.

Yu, Y., El Kamel, A. and Gong, G. (2013b). Modeling intelligent vehicle agent in virtual reality traffic simulation system. 2013 2nd International Conference on Systems and Computer Science (ICSCS), pp. 274–279, Villeneuve d'Ascq, France: IEEE.

Chapter **10**

The Sense of Touch as the Last Frontier in Virtual Reality Technology

Jonatan Martínez, Arturo S García, Miguel Oliver,
Pascual González and *José P Molina**

1. Introduction

In recent years, a renewed interest in virtual reality has been witnessed, thanks to new headsets that offer very high resolution and wide field of view, coupled with room-scale tracking and affordable prices for the mainstream consumer. In this way, everybody can now experience virtual worlds with crisp and colorful images. However, if the user reaches out his hand to touch and feel a virtual object, the experience then turns blurry. At the moment of writing this chapter, touch is still the last frontier of virtual reality, and it will remain as such for years to come. The reason for that is, on the one hand, the complexity of human sense of touch and, on other hand, that there is no a winner technology, Every solution has its own pros and cons which one has to know and assess.

A decade ago, the LoUISE research group challenged the problem of creating an affordable touch glove, and after several years of work, this ended with a vibro-tactile platform, called VITAKI (VIbroTActile prototyping toolkIt) (Martínez et al., 2014), which easily solves the problem of creating any touch device, from gloves to bodysuits, and for many applications, from games to rehabilitation systems. This chapter details how and why the research team made their mind up for vibration technology, in particular Eccentric Rotating Mass (ERM), instead of other solutions that are also cited here. As it has already been proved in previous publications, this vibro-tactile technology can be used to feel and follow the shape and surface of a virtual object; even assess its size and weight, and it is being used in new projects to explore how far we can go with it.

This chapter begins by offering an overview of the physiology of touch and some of the parameters that have been studied over time to design haptic devices. It is followed

Instituto de Investigación en Informática de Albacete, Universidad de Castilla-La Mancha, Spain.
* Corresponding author: JosePascual.Molina@uclm.es

by a comprehensive review of the different technologies used by other researchers, and the choice of selected actuators is argued. After that, it focuses on the work related to the technology that is used at the LoUISE lab, and at the end of this chapter, some conclusions are drawn and discussed.

2. Physiology of Touch

The sense of touch is the one that occupies the largest area in the human body and is formed by two main sensory systems: the kinesthetic system, which perceives the sensations produced in muscles, tendons and joints, such as that caused by movement; and the cutaneous or tactile system, which responds to stimuli from the surface of the skin. These stimuli can be thermal, electrical, chemical, pain or deformation of the skin, as explained in the next sections.

2.1 Mechanoreceptors

A mechanoreceptor is a type of sensory receiver that responds to mechanical pressures or distortions. When the skin is subjected to pressure or vibration, the skin surface is distorted, generating waves that are transmitted through the skin, reaching the membranes of the mechanoreceptors. The membrane of these sensors is also altered, causing ion channels to open and in turn, an alteration in the electrical potential that is transmitted to the sensory cortex, produces a tactile sensation, depending on the type of activated receptor.

In the human being, the glabrous skin (without hair) of the hand has a great density of these elements, with approximately 17,000 nerve endings that are concentrated mainly in the fingertips (Vallbo and Johansson, 1984). They are normally divided into four main types, whose representation can be seen in Fig. 1.

- Ruffini's endings (corpuscles): They are distributed in the deep zone of the dermis, with a low spatial resolution. They are sensitive to sustained pressure and lateral deformation of the skin. They intervene mainly in the perception of continuous stimuli, the detection of the direction of movement in lateral stimuli of the skin, and in the proprioception of the position of the fingers.

- Merkel's disks: They are found in the epidermis with a great spatial density. They are sensitive to sustained pressure at very low frequencies (less than 5 Hz), and to spatial deformation. Their main function is detection of low-intensity frequencies, rough perception of textures and shape detection.

- Meissner's corpuscles: They have a high spatial resolution and are distributed just below the epidermis, in a very shallow area. They are sensitive to temporary changes in the deformation of the skin (between 5–40 Hz) and to spatial deformation. Their function is detection of low-frequency vibrations.

- Pacinian corpuscles: They are located in the deep dermis and, due to the large size of their receptor field, provide a low spatial resolution. They are sensitive to changes in the deformation of the skin at high frequencies (40–500 Hz). So, in addition to detecting vibrations, they are used for the fine perception of textures.

Table 1 presents a summary of these characteristics. Alternatively, some authors appoint these recipients, based on their speed of adaptation. Thus, they are classified

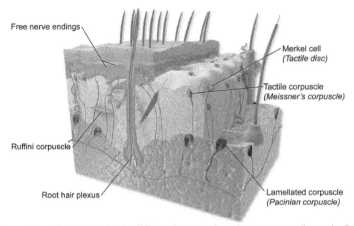

Fig. 1: Cross-section of skin, showing its different layers and mechanoreceptors (image by Bruce Blaus).

Table 1: Four-channel mechanoreception model (Table Adapted from Johnson, 2002).

Nerve Ending	Ruffini's Endings	Merkel's Disks	Meissner's Corpuscles	Pacinian Corpuscles
Type	SAII	SAI	FAI	FAII
Velocity	Slow	Slow	Fast	Fast
Size	Big	Small	Small	Big
Location	Subcutaneous	Superficial	Superficial	Subcutaneous
Frequency	Static	0–100 Hz	1–300 Hz	10–1000 Hz
Peak sensitivity	0.5 Hz	5 Hz	50 Hz	200 Hz
Resolution	> 7 mm	0.5 mm	3 mm	> 10 mm

as Slow Adapting receivers (SAI and SAII) and Fast Adapting (FAI and FAII), with the correspondence shown in Table 1. It is important to note that a sensation of pressure may come from the activation of several specialized mechanoreceptors at the same time, instead of just one. Therefore, all of them are necessary to allow the manipulation and grip of objects in a stable and precise way.

2.2 Distribution

As described above, the tactile stimulus receptors are not encapsulated in a single organ but are distributed throughout the body. However, this concentration is not uniform. Certain parts of the body, such as the hands and, in particular, the fingertips, have a greater sensitivity to external stimuli, as can be seen in Fig. 2. This unequal distribution indicates the most propitious places to transmit tactile information.

2.3 Detection Threshold

The threshold of detection is the minimum intensity of a signal to be detected by human sense of touch. In the case of mechanical stimuli, the minimum detectable intensity is measured by the amplitude of the movement at a certain frequency.

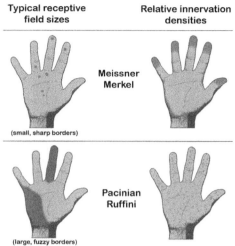

Fig. 2: Distribution of sensory receptors in the hand (adapted from Vallbo and Johansson, 1984).

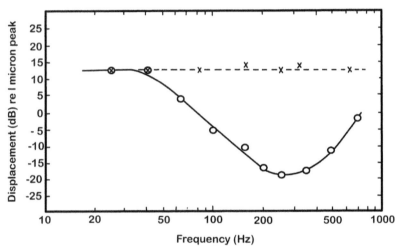

Fig. 3: Curve of the sensitivity threshold of the skin of the hand (adapted from Verrillo, 1975).

Thus, the minimum threshold is found in frequencies around 250 Hz (threshold of ~ 0.0001 mm), where the maximum sensitivity peak of the mechanoreceptors is found (Verrillo, 1962). Figure 3 shows the curve of the detection threshold of the skin of the hand.

2.4 Perception of Intensity and Level of Discomfort

Different researchers in the field of psychophysics have studied the subjective perception of the intensity of vibration (Stevens, 1959; Verrillo et al., 1969), demonstrating that the relationship between the amplitude of vibration and the perceived magnitude follows a power function. They also discovered that for the same level of vibration, women perceive it as subjectively more intense than men and that, as a person ages, he gradually loses his sensitivity. Knowing the subjective perception function for a given actuator is useful in order to provide the user with the desired intensity values (Murray et al., 2003; Ryu et al., 2010).

Finally, it is important to maintain an adequate level of intensity to avoid discomfort in the user, or problems, such as the well-known Raynaud syndrome or the white finger (Bovenzi, 1998). This occurs in extreme cases where a worker is exposed to strong vibrations produced by machinery for a long time, which damage the circulatory system.

The magnitude of discomfort depends on many factors, such as frequency, direction of vibration or exposure time. Griffin (1996) collects several studies in which semantic scales are established to define the intensity of vibrations perceived by users. In these studies, it can be observed that, in general, vibrations above 0.7 $m \cdot s^{-2}$ are considered as not pleasant.

3. Touch Feedback Technologies

The transmission of a tactile stimulus is achieved through touch displays, that is, devices that, in contact with the skin, are capable of generating sensations of greater or lesser complexity. The most common displays are composed of a single actuator that produces a tactile sensation, such as pressure. The most complex, however, are formed by a matrix of mobile pins forming a surface, in such a way that they can move independently to make them stand at different levels and represent textures or shapes. The main characteristics that differentiate tactile displays from others are their spatial resolution, that is, how many individual elements make up the display per unit area, temporal resolution, response time or frequency response, and the strength or intensity that they can exercise.

The most decisive factor in terms of the variety of sensations that can be transmitted with these displays is the technology used in it. Currently, a wide variety of technologies have been used, such as pneumatics, electromagnetic, electrostatic, piezoelectric, alloys with memory thermal effect as well as other less common ones, such as electrotactile stimulation, Functional Neuromuscular Stimulation (FNS) or electroreological fluids. Next, each of them will be detailed and examples proposed in the literature will be discussed.

3.1 *Pneumatic Actuators*

The pneumatic actuators are characterized by using compressed air to expel it directly on the skin by means of microinjectors, or to fill small bags that cause pressure on the skin when inflating. Pneumatic touch displays can be thin, lightweight and flexible since the device that creates the air pressure can be mounted away from the actuator itself, although this makes it less portable. Traditionally, a compressor, an accumulator and electrovalves, that allow the passage of air through the different conduits to the actuators, have formed this system. Another alternative is the use of pistons actuated by electromagnetic solenoids, and more recently, the use of metal hydride (MH) alloys is being investigated. These alloys are characterized by capturing hydrogen when they are cooled and released, when heated. By joining them to a Peltier device, capable of varying its temperature by means of an electric current, there is a portable pressure variator for pneumatic devices (Sato, 2007).

This type of technology can create tactile sensations of great intensity. However, the main disadvantage of the pneumatic displays lies in their operating frequency, which is around 10 Hz due to the compressibility of air. Below are some of the devices created through this technology.

The research group led by Shuichi Ino (Sato, 2007) has carried out several research works on tactile interfaces based on pneumatic actuators. In particular, they have investigated the displacement and pressure sensations created from pneumatic actuators capable of moving about 3 mm laterally and creating pressures of 5.9 N.

Teletact, created by UK's National Advanced Robotics Research Center (ARRC) and Airmuscle Limited (Stone, 2000), is a glove based on pneumatic technology that includes a pump, tank and pressure control channels to inflate small airbags that can be mounted on data gloves. The first prototype used 20 pneumatic bags that could be inflated to 13 psi. The second revision of the glove, called Teletact II and shown in Fig. 4a, had 30 airbags with two different pressure ranges. One of them is located in the palmar zone and supports a maximum pressure of 30 psi, compared to the 15 psi of the rest.

Using the same airbag technology, the ARRC team has also created the Teletact Commander (Stone, 2000), a multifunction hand controller that has three built-in pneumatic actuators on its surface that can be controlled by a compressor or by a piston actuated by a solenoid. This device can be seen in Fig. 4b.

Finally, King et al. (2008) integrated multiple pneumatic elements to form a 3×2 elements display. This device has dimensions of 10×18 mm and each element is a small 3 mm silicone balloon. It has four bits of resolution, a maximum pressure per element of 0.34 N and a frequency response of 7 Hz.

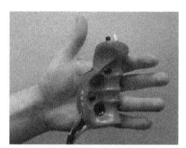

(a) Teletact II   (b) TeleTact Commander

Fig. 4: Teletact pneumatic actuators (Stone, 2000).

3.2 Piezoelectric Actuators

Piezoelectricity is a phenomenon that occurs in certain crystals by which they deform when subjected to an electric field. Piezoelectric actuators are easy to find commercially, small, flexible and thin. They can generate large forces in a wide range of frequencies, but the displacement they generate is low, around 0.2 per cent. So they are usually mounted with lever mechanisms, in multiple layers, or bimorphically. A bimorph actuator consists of two long, joined, piezoelectric elements.

When a voltage is applied, one of them shrinks and the other stretches, getting the bimorph to bend. A general disadvantage of piezoelectric actuators is that a very elaborate system is needed to mount them, in addition to the use of high voltages, which can also affect safety (Edmison et al., 2002).

Numerous touch displays based on piezoelectric elements have been built. Debus et al. (2002) presented the design and construction of a multi-channel vibro-tactile display composed of a handle with four piezoelectric actuators to transmit stimuli in four directions. Summers et al. (2001) constructed a device consisting of a matrix of 100×100 piezoceramic tactile elements, with a spatial resolution of 1 mm to study the accuracy of perceived sensations. His study determined that at a frequency of 320 Hz, spatial accuracy is greater than at 40 Hz.

STRESS, created by Pasquero and Hayward (2003), is based on bimorphic piezoelectric elements, and allows the reproduction of sequences of tactile images at about 700 Hz. The display uses a set of 100 tactors that move laterally to the surface of the skin. The density is one contactor per square millimeter, so it has high temporal and spatial resolution.

A new version, presented in Wang and Hayward (2005), consists of 6×10 actuators, a spatial resolution of 1.8×1.2 mm, a maximum deflection of 0.1 mm and a frequency of about 250 Hz (Fig. 5a).

Kyung et al. (2004) designed and built a tactile display based also on bimorphic piezoelectric actuators attached to a 2D mouse-shaped device, which also provides kinesthetic force and skin-moving sensation. The touch screen is composed of eight elements with each moving five pins at a maximum of 1 mm, with a force of 1 N and a maximum frequency of 1 kHz (Fig. 5d). This group also created a haptic display, consisting of 30 actuators of the same type that powered a 5×6-pin matrix (Kyung et al., 2005). The spatial resolution is 1.8 mm, with a response time of 500 Hz and a displacement of 0.7 mm.

Zimmerman and his group added piezoceramic actuators to a VPL DataGlove (Zimmerman et al., 1987). They used frequency modulation to vary the intensity of the tactile sensation and to minimize the sensation of numbness.

(a) Stress V2 (Pasquero and Hayward, 2003)   (b) Bimorph actuator SP-21b (Mide Corp.)

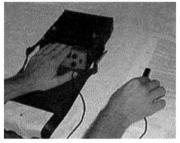

(c) Optacon (Bliss, 1969)   (d) Integrated Tactile Display (Kyung et al., 2004)

Fig. 5: Haptic devices based on piezoelectricity.

The Optacon (Bliss, 1969) was a commercial device formed by a matrix of 24 × 6 pins connected to bimorphic piezoelectric actuators and a small handheld camera that allowed to read printed text to people with visual disabilities. The pins vibrate at a fixed frequency of 250 Hz (Fig. 5c).

Recently, the company Mide (http://www.mide.com) has put on sale a piezoelectric kit consisting of a bimorph actuator (Fig. 5b) and a two-channel controller for $ 300.

3.3 Thermal Actuators

Thermal actuators are based on materials that can vary their temperature in the presence of electric currents to transmit tactile sensations. This type of actuator has the disadvantage of its low response speed. In addition, it can be dangerous for the user in the event of a failure of the temperature sensor or the control loop. The Displaced Temperature Sensing System, which was commercially available and was developed by C & M Research (Burdea and Coiffet, 2003), provides thermal stimulation in the fingers. Each of the actuators is composed of a combination of thermoelectric heat pump, temperature sensor, and heatsink. This allows the actuator to obtain feedback from the sensor in order to regulate the surface temperature to the desired value. This can vary between 10–45°C with a resolution of 0.1°C. The X/10 model supports up to eight channels, and the actuators can be in the form of Velcro strips or thimbles.

Ino et al. (1993), developed a tactile interface composed of a Peltier module, which is the thermal actuator itself, and a thermocouple, which acts as a temperature sensor. By means of this device, they pursue that the user can distinguish, in a virtual environment, materials with different degrees of thermal conductivity like, for example, wood and aluminum.

3.4 Alloys with Thermal Memory Effect (SMA)

Alloys that show memory a thermal effect (Shape Memory Alloys, SMA) are characterized by changing shape at low temperatures and recovering their initial state when heated. This is achieved by passing a large current through wires of this material, so they have a high consumption. They can exert great forces with a displacement of between 2–4.5 per cent. Due to the thermal capacity of the material, changes in temperature take time and actuators based on SMA usually do not exceed 10 Hz. In addition, care must be taken to adequately isolate the skin to avoid injury from burns.

One of the main applications of this technology is the creation of displays in the form of a rod matrix. Johnson (1990) proposed a matrix of 5 × 6 pins separated by 3 mm, with a response time of 100 ms, 1 second of recovery (cooling of the alloy), and a maximum force of 0.196 N.

The display of Wellman et al. (1998), uses liquid cooling to achieve shorter recovery times, thus achieving a frequency of up to 40 Hz. It is composed of ten pins spaced 2 mm and pushed by an SMA wire with V configuration that exerts a maximum force of 1.5 N (Fig. 6a). Other similar devices can be found in Hasser (1993), Taylor et al. (1997), Kontarinis et al. (1995).

Scheibe et al. (2007) presented a device formed by SMA cables of about 50 mm adjusted around thimbles (Fig. 6c) so that they contract around 1.5–2.5 mm, exerting pressure on the fingertip. The cable used is 80 μm, which provides a response time of less than 50 ms.

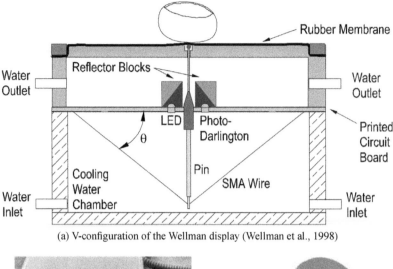

(a) V-configuration of the Wellman display (Wellman et al., 1998)

(b) Tactool System, by Xtensory   (c) SMA thimbles of Scheibe et al., 2007

Fig. 6: Haptic devices based on alloys with thermal memory effect.

Tactool System is a commercial Xtensory product whose touch displays are mounted on the fingertips by Velcro strips (Fig. 6b). The tactors are based on pins operated by SMA, and can be used impulsively (30 g) or vibration (20 Hz). Despite not being a widespread technology, it is now possible to find it commercially. The company Mide (http://www.mide.com), for example, offers a kit for about $ 500, consisting of sheets and SMA wire.

3.5 *Microelectromechanical Actuators (MEMS)*

Microelectromechanical systems (MEMS) are microscopic mechanical systems coupled to electrical or electronic circuits (Mahmoudian et al., 2004). MEMS have been used to a large extent as accelerometers, gyroscopes, high quality oscillators, microphones and amplifiers, among other things.

Some authors, such as Enikov et al. (2002), have applied this technology to build touch displays. In this case, the device consists of 4 × 4 actuators matrix that vibrates, thanks to piezoelectric elements. The matrix uses MEMS technology to create micro-brakes, and each individual actuator is activated and deactivated by pairs of thermoelectric actuators. The display of Streque et al. (2008) also has the form of an array of actuators, but in this case actuated by miniature electromagnetic coils and neodymium magnets, forming a matrix of 4 × 4 elements with a resolution of 2 mm.

3.6 Electromagnetic and Neuromuscular Interfaces

Electromagnetic interfaces use electrodes to pass a current through the skin. Neuromuscular interfaces use electrodes directly under the skin to achieve muscle stimulation.

They have not been used on a large scale because of their invasive and dangerous nature, since for the user the pain frontier is very close. Other drawbacks are the low spatial resolution that can be detected, and the instability in the relationship between electric current and perceived sensation.

People can handle sharp objects, such as knives or needles, in everyday life although they can cause pain because it depends on the force with which they are touched, the tactile sensation being very weak if the force with which the person touches them is also weak.

In electro-tactile actuators, this is not the case, since the sensation when touching an electrode is independent of the force with which it is touched, which can cause some fear and rejection. To avoid this effect Kajimoto et al. (2001) used electrotactic interfaces attached to a pressure sensor, so that the intensity of the sensation was proportional to the force exerted. These researchers try to avoid the problem of resolution using anodic current, instead of cathodic, so that in this way, the vertically-oriented nerves are stimulated, and the sensation is more localized, creating in the user the perception of a vibration. The prototype they build is composed of a matrix of 2 × 5 electrodes and a spatial resolution of 2.54 mm (Fig. 7a). The pulse duration is set between 0–0.5 ms with an amplitude between 0–10 mA.

SmartTouch, also created by Kajimoto et al. (2004), uses an optical sensor and an actuator formed by a 4 × 4 array of 1 mm diameter electrodes (Fig. 7b). Its purpose is to capture a visual image and show it through electric stimuli of 0.2 ms, 100 to 300 volts and 1–3 mA. In this way the user can notice, for example, the Braille signs printed on a paper.

 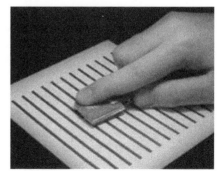

(a) Electrode array of Kajimoto et al. (2001)   (b) SmartTouch (Kajimoto et al., 2004)

Fig. 7: Haptic devices based on electrotactile actuators.

3.7 Electrostatic and Electrostrictive Materials

Electrostatic actuators use Coulomb forces generated by the electric field between two charged surfaces. This requires very high voltages and, even then, the force and displacement are very small.

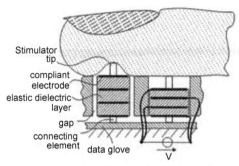
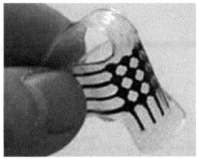

(a) Structure of the electrostatic stimulator with an elastic dielectric (Jungmann and Schlaak, 2002)

(b) Wearable tactile display formed by electroactive polymers (Koo et al., 2008)

Fig. 8: Haptic devices based on electrostatic and electrostrictive materials.

Jungmann and Schlaak (2002) created a miniaturized actuator, using multiple layers of elastic dielectrics that are compressed when a voltage between 100–1000 V is applied (Fig. 8a). The device is flexible, lightweight and low cost, which makes it suitable for use, for example, in data gloves or with return of forces.

Tang and Beebe (1998) designed an electrostatic tactile display consisting of a set of metal electrodes, covered by a thin insulating layer. As a voltage is applied between the metal electrode and a human finger touching the insulating layer, an electrostatic force attracts the skin of the finger. While moving the finger along the surface of the display, the frictional forces will vary according to the electrical potential of the electrodes. This principle allows very fine and adequate touch displays with a high actuator density. The problems of this system are that, on the one hand, the forces are extremely small and, on the other, a relative movement between the skin and the display is necessary.

Electrostrictive materials are dielectrics that change their shape when they are under the effect of an electric field. An example is the so-called electroactive polymers used by Koo et al. (2008) to create a tactile display whose membranes are warped when applying high voltage (Fig. 8b). Their display is formed by a 4 × 5 elements matrix with a maximum displacement of 0.9 mm at 3 kV, a force of 14 mN and a weight of 2 gr.

3.8 Electrorheological and Magnetorheological Fluids

Electrorheological fluids (Electro Rheological Fluid, ERF) and magnetorheological (Magneto Rheological Fluid, MRF) can vary their viscosity significantly in the presence of an electric field (ERF) or an electromagnetic field (MRF). This change, which occurs in a millisecond response time, is reversible and can be applied, for example, to displays where the user examines a tactile graphic by moving his finger across its surface. As these devices cannot exert active forces, they are not suitable for displays with tactile feedback in telemanipulation systems. They also require high voltage and are incapable of representing very rigid surfaces or well-defined edges.

Kenaley and Cutkosky (1989) were the first to create a tactile sensor based on ERF in the form of a thimble for robotic fingers. They also proposed an actuator based on the same technology composed of 4 × 3 cells (Voyles et al., 1989). Monkman (1992) proposed the application of ERF for touch displays. Taylor et al. (1997) presented improvements to previous displays based on ERF using a layer of cloth inside the fluid layer, thus achieving duplicating the reactive forces with less current and improving safety, since it isolates the electrodes. Its display is composed of 5 × 5 touch units of 11 mm on each side and separated by 2 mm each. The applied voltage is quite high, about 3 kV. Liu et al. (2005) used instead magnetorheological fluid to create a display with a single cell and testing two different types of magnets. They concluded that the MRFs are also suitable for this type of passive displays.

Klein et al. (2005) made tests with this type of fluids to create a prototype of a 3D touch screen formed by microcells, to apply it to medicine. Voyles et al. (1996) created both a sensor and a thimble-shaped actuator using ERF to observe human tasks that require contact. Finally, the MEMICA system (Bar-Cohen et al., 2001), developed by Rutgers University, uses gloves with haptic feedback based on ERF.

Kim et al. (2009) used magnetorheological fluid to create a manipulator for the hand, based on passive actuators.

3.9 Technologies Without Direct Contact

The emergence of user interfaces based on gestures has led to the introduction of technologies capable of stimulating touch through the air. Aireal (Sodhi et al., 2013), for example, is based on the generation of air vortices created by a mobile nozzle that can direct them towards the user. Combined with depth cameras to locate the user's hand in space, it can create small impulses in the skin. The device has an approximate range of one meter and a latency of about 140 ms.

On the other hand, UltraHaptics (Carter et al., 2013) used an array of 16 × 20 ultrasound transducers to create multiple feedback points in the air. To do this, it uses the principle of acoustic radiation force, which is created when a set of actuators emit a frequency in phase. The system is capable of producing tactile points in space, which are perceived by the user as a slight tingling on the surface of the skin.

This type of technology, although it may be interesting in certain very specific areas, is very limited. In the first case, the air vortices are discrete impulses that are released to the user, and in addition to having a high latency, are very scattered stimuli that affect a large area of the skin. In the second case, the workspace is very limited, since it is determined by the size of the array of actuators. In addition, the sensation is very subtle, and could easily go unnoticed by the user.

3.10 Electromagnetic Actuators

3.10.1 Solenoids and Mobile Coils

A copper coil, that creates an electromagnetic field when it is crossed by an electric current, forms solenoids. This phenomenon is used to attract a ferromagnetic core and thus produce a mechanical movement capable of exerting pressure or vibration. The mobile coils follow a similar principle, but consist of a permanent magnet and it is the coil that moves by the electromagnetic effect. Unlike the solenoids, they have a linear response and are bidirectional (they can exert force in both directions), while the solenoids need a spring to return to the initial position. There are other variations in which the coil is fixed and the magnet is mobile, which have a similar response.

These devices are cheap and easy to control, but are usually relatively bulky. A great limitation of the actuators based on small size solenoids is that the force they can exert is very limited; so it is most feasible to use them to transmit vibration sensations.

This technology is present in many prototypes. One of them is the haptic BubbleWrap display (Bau et al., 2009), shown in Fig. 9a. It consists of a set of cells composed of a flat spiral-shaped copper solenoid attached to a permanent flat magnet. These cells can contract and expand individually about 10 mm to create both haptic return by vibration, and haptic passive return, adopting different forms and degrees of firmness. Kontarinis et al. (1995) created a tactile display, modifying small 0.2 W speakers that joined small manipulators for two fingers. The characteristics of the obtained device are a range of movement of 3 mm, and a maximum force of 0.25 N at 250 Hz. ComTouch (Chang et al., 2002) is a communication device that converts in real time the pressure of the hand exerted by a user, in intensity of vibration that another remote user receives. In this way, voice communication is enriched by complementing it with a touch channel. They use small speakers (V1220, from AudioLogic Engineering).

Some authors have also created matrices based on solenoids. Fukuda et al. (1997) proposed a matrix of actuators composed of a micro coil and a permanent magnet that moves inside it. Each actuator is 2 mm in diameter and exerts a force of 7.6 mN/mm^2, which it uses to produce vibro-tactile sensations of displacement in the skin, and thus be able to study tactile parameters. Petriu and McMath (1992) describe another of these tactile devices created for tele-manipulation and formed by 8 × 8 pins on a surface of 6.5 cm^2. Talbi et al. (2006) used small solenoids to create a 4 × 4 vibro-tactile pins matrix with an oscillation frequency of 250 Hz, an amplitude of 0.2 mm and a force of 1.2 mN. The scheme of the actuators is shown in Fig. 9d. Another similar device, called VITAL and created by Benalikhoudja et al. (2007), is also based on microsolenoids and has

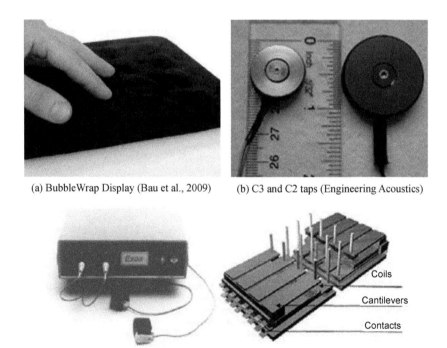

(a) BubbleWrap Display (Bau et al., 2009)   (b) C3 and C2 taps (Engineering Acoustics)

(c) Exos Touchmaster   (d) Vibro-tactile pin matrix by Talbi et al. (2006)

Fig. 9: Haptic devices based on solenoids and mobile coils.

a resolution of 4 mm, a force of 13 mN, maximum deflection of pins of 0.1 mm and frequencies of up to 800 Hz. Tan and Pentland (1997) created a touch screen consisting of a 3 × 3 vibro-tactile stimulators matrix placed on the forearm and separated by 8 cm to show directional patterns.

There are also commercial actuators, such as the range of Engineering Acoustics (http://www.atactech.com), with a price of about 200 dollars and composed of taps C2, C3, CLF and EMS. Figure 9b shows two of these tactors. The C2 tactor is quite common in the field of research and has been used, for example, by Gurari et al. (2009) to simulate the hardness of a material in a virtual environment and compare the effect when the foot, arm or fingertip is stimulated. The company EXOS launched TouchMaster, which consists of mobile coil actuators for the four fingertips that are adjusted by Velcro strips (Fig. 9c). The return that is obtained is of vibro-tactile type, with a fixed response frequency around 210–240 Hz. Optionally, electronics can also be added to vary the amplitude and frequency. Audiological Engineering created several models of a transducer, called Tactaid, which, although no longer available, have been used by several researchers (Ryu et al., 2010; Chang et al., 2002; Hein and Brell, 2007).

3.10.2 Linear Resonant Actuators (LRA)

A fixed coil and a magnet fixed to a mass form the Linear Resonant Actuators (LRAs). The magnet and the mass, whose clamping depends on a spring, are attracted and repelled by the coil producing an oscillating movement. Due to the inclusion of this mass, the system has much more inertia than the systems described above; so it can only oscillate effectively at its resonant frequency. Recently they are being included in more and more mobile devices due to their low consumption and high intensity of vibration; so it is easy to find them in the market at a reduced price. Precision Microdrives, for example, has a model (C10-100) with a vibration amplitude of 1.4 G, resonant frequency of 175 Hz, and consumption of 69 mA for about 10 euros.

They have also been used in research in recent years. For example, Seo and Choi (2010) used them in a study to create illusions of vibro-tactile linear movement by means of two LRAs, integrated in a rigid support.

3.10.3 Engines and Servomotors

An electric motor is a device made up of coils and a rotor that rotates when an electric current is applied to it. A servomotor is a motor that has the ability to be located in any position within its operating range and to remain stable in that position, thanks to a feedback loop. This rotational movement can be used to create tactile sensations through different types of mechanisms.

The device of Shinohara et al. (1998) is a display that shows tactile graphics. It is composed of a matrix of 64 × 64 pins with a separation of 3 mm. The extension of the stimulators is done by axes that are controlled by step-by-step micro-motors. The maximum height of the high stimulators is 10 mm with a resolution of 1 mm, but the forces generated are too small to extend the stimulators while they are touched by the user. The refresh time is 15 s—too high for the recognition of virtual objects in real time, but appropriate for static objects.

Ottermo et al. (2005) used small brushless motors to rotate a screw that moves each pin lengthwise. Its display consists of 4 × 8 elements that move a maximum of 3 mm generating a maximum force of 1.7 N, with a resolution of 2.7 mm and a frequency of 2 Hz.

Wagner et al. (2004) improved this response time by achieving a refresh of up to 7.5 Hz, using servomotors to create a 6 × 6-pin display, with 2 mm separation and 2 mm resolution (Fig. 10a).

Because the set of servomotors is usually bulky, several attempts have been made to separate them from the tactile zone. For example, Sarakoglou et al. (2005) used nylon threads and springs to connect small electric motors to a 4 × 4-pin display. These pins have a maximum displacement of 2.5 mm, resolution of 2 mm, a maximum force of 1 N per pin and maximum frequency of 15 Hz.

Another very different concept is the Inaba and Fujita (2006) actuator, which uses a motor to wind a strip of cloth that exerts pressure on the fingers. Minamizawa et al. (2007) extended this concept by designing a device, called Gravity Grabber with two motors, so that this strip of cloth can also exert displacement sensations on the skin (Fig. 10b). These sensations of displacement are very useful in areas, such as remote manipulation, where they help determine the correct degree of force that the fingers must exert on an object so that it does not fall. To achieve this purpose, Chen and Marcus (1994) did not use a strip of cloth in contact with the skin, but directly the axis of an engine, thus obtaining one degree of freedom. Webster et al. (2005) used a small sphere in contact with the user's finger which, joined to two perpendicular motors, allows it to rotate with two degrees of

(a) Display by Wagner et al. (2004)   (b) Gravity Grabber by Minamizawa et al. (2007)

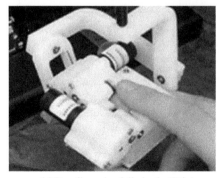

(c) Two-dimensional display by Webster   (d) Interface composed of Frati and Prattichizzo
et al. (2005)   (2011) motors

Fig. 10: Motor-based devices.

freedom (Fig. 10c). Finally, the display of Frati and Prattichizzo (2011) is based on three motors that wind a thread, each of them stretching the tip of a rigid piece placed under the finger, thereby reflecting the contact with surfaces of different inclination (Fig. 10d).

3.10.4 *Eccentric Rotating Mass (ERM) Actuators*

Eccentric Rotating Mass (ERM) actuators or vibrating motors are small electric motors whose shaft is coupled with an off-center mass, making the assembly vibrate when it is rotated. Thanks to its wide use in mobile devices, vibrating motors have been favored by a great advance, being smaller, efficient, and cheap. Commercially, two types can be found: cylindrical and flat. They are easy to use, since they only need a small direct voltage between their terminals, still providing a considerable vibration force. Its low cost and effectiveness have made it widely used in all types of devices, from game-oriented systems, such as joysticks, gamepads or steering wheels, to mobiles or PDAs.

In the field of virtual reality, they have also been well received, and can be found commercially, for example, in the CyberTouch glove, released by CyberGlove Systems. Basically, it is a CyberGlove data glove, capable of measuring the position of the fingers, in which a vibro-tactile stimulator has been added to each finger, and one more in the palm of the hand (Fig. 11a). These are located in the dorsal part, and are of cylindrical type. Its frequency can be varied from 0–125 Hz and the maximum amplitude is 1.2 N to 125 Hz.

These actuators have also been included in numerous wearable designs, that is, integrated into conventional clothing. An example is the shoulder-shaped device of

(a) CyberTouch vibro-tactile glove

(b) ActiveBelt vibro-tactile belt, Tsukada and Yasumura (2004)

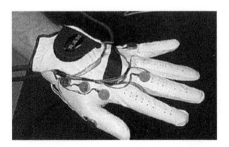

(c) Vibro-tactile glove, Zelek et al. (2003)

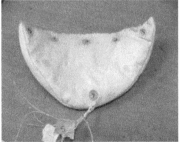

(d) Vibro-tactile shoulder pad, Toney et al. (2003)

Fig. 11: Devices with ERM actuators.

Toney et al. (2003), where authors use flat vibrators to investigate the use of vibro-tactile displays in tissues (Fig. 11d). Another design presented by Tsukada and Yasumura (2004) and named ActiveBelt, consists of a belt with eight vibro-tactile actuators integrated and which, combined with a GPS, allow the user to obtain directional information through touch (Fig. 11b). Tacticycle, created by Poppinga et al. (2009), also transmits directional information, but in this case the system is composed of two vibrators placed on each part of the handlebar of a bicycle.

The tactile device created by Zelek et al. (2003) is more complex and tries to represent visual information through a vibro-tactile system. To do this, they create a description of the scene and its certainty by means of a system of stereo cameras that map the surrounding environment to a glove with 14 vibrators placed on its back (Fig. 11c). The arrangement of these vibrators is limited by the size of the skin fields receptive to vibrations, so arranged that each motor can be univocally identified by the user when several motors are activated simultaneously.

3.11 Discussion

The main consideration when selecting the tactile actuators for the VITAKI platform was the ability to find sufficiently light and small devices that can be adapted to the body of a user so they can, for example, be part of a glove. In addition, it is important that they can be found commercially, ideally at a low price.

None of the displays described satisfy all the requirements to be an ideal touch display, such as cost, volume, complexity or resolution. In addition, most have other disadvantages, such as their high weight, large volume or structural rigidity that makes them unsuitable to be wearable and introduce them, for example, in data gloves. Finally, only some technologies are commercially available, which are compared below. Table 2 shows some of their main characteristics.

- *Piezoelectric Actuators*: Although piezoelectric technology is present in a large number of consumer devices, such as loudspeakers and cigarette lighters, piezoelectric switches have been used almost exclusively by researchers. Its biggest advantage is its reduced thickness and large bandwidth, facilitating oscillations at frequencies between 1–300 Hz. However, its limited representation in the market means that the few models that can be found are not suitable for all possible applications. In addition, they are expensive, fragile, and operate with high voltage (~ 100 V). So it is essential to encapsulate them and ensure their electrical isolation to be safe for the user.

Table 2: Comparison of different tactile actuator commercial technologies.

	LRA	ERM	Piezoelectric Actuators	SMA	Mobile Coils
Size	Small	Small	Variable	Variable	Big
Thickness	Thin	Thin	Minimum	Variable	Large
Latency (ms)	20–30	40–80	< 1	High	< 1
Frequency (Hz)	~ 175	50–250	1–300	~ 10	200–300
Voltage (V)	< 5	< 5	50–200	< 5	< 3
Price (€)	5–10	1–5	50–170	Variable	~ 200

- *Alloys with Thermal Memory Effect (SMA)*: Its use as a touch transducer is even smaller than piezoelectric actuators, and it is only in scientific literature that some prototypes can be found. Although SMA wires can be found commercially, it is necessary to give them an adequate form to serve as tactile actuators. In addition, because its shape depends on changes in temperature, its thermal inertia prevents them from being very fast. So they have a very poor frequency response.

- *Mobile Coils*: They have a good frequency response and, although they are not very common outside the scope of the research, they are commercially available mainly for military, medical, or aviation purposes. Their main disadvantage is the high price per unit and their size, which makes it unsuitable to integrate them into small spaces, such as the fingertips.

- *Linear Resonant Actuators (LRA)*: With a competitive price per unit and a small size, they have characteristics similar to ERM. Other advantages are low consumption and great durability, which makes them ideal for mobile devices. However, they can only work at their resonant frequency (about 180 Hz), which means that only an amplitude modulation can be made and that a specialized circuit is needed to find the resonant frequency at each moment, which changes according to external factors, such as assembly.

- *Vibrating Motors (ERM)*: These are the actuators that most of the industry of mobile devices and video games has adopted; so their price is very low and can be found easily. They are robust, easy to operate, safe for the user and small, and yet are able to provide a high intensity of vibration. Among their disadvantages are their relatively high latency for being an inertial device, and the impossibility of modulating separately the amplitude and frequency of vibration, since both depend on the applied voltage.

In view of the characteristics of the different available technologies, the choice for this research work has been the use of ERM actuators. Another reason that justifies this choice is the Brown and Kaaresoja (2006) experiment, in which the rhythm and intensity of the vibration of ERM vibrators and commercial C2 tactors (composed of an electromagnetic coil) are varied. The results showed little difference when recognizing the vibro-tactile signals using these two systems. This suggests that, in many cases, the choice of ERM actuators may be more appropriate than other tactors with more bandwidth but much more expensive and bulkier.

Hardware controllers used by researchers to connect ERM actuators usually have, however, remarkable flaws. The controller used by Bloomfield and Badler (2007), for example, is composed of several relays. These electromechanical devices are only able to completely switch on or off the vibrators, so their intensity cannot be adjusted. Another approximation found is the use of a digital/analog converter chip to modulate the vibration intensity (Tsukada and Yasumura, 2004), connecting the motors directly to the output. However, the circuit used provides a maximum of 3 mA of current, whereas this kind of actuator typically requires more than 50 mA. Some solutions, as the used by Ferscha et al. (2008), present several limitations. In this case, the intensity level can be adjusted by PWM, but it affects globally all the channels. Sziebig et al. (2009) performed software PWM, which produces an unnecessary overload in the system, achieving a reduced frequency and resolution in comparison with the hardware version.

To improve the response of this kind of actuators, the VITAKI platform includes two well-known driving schemes—the first one, *overdrive*, consists of generating a short pulse of relatively high voltage to reduce the starting delay of the motors, and the second one, *active braking*, stops them using a short pulse of reversed voltage. In addition, this tool includes a software module to facilitate the specification of the vibratory stimulus associated. This software presents some novel features, like the ability to work with overdrive and negative ranges and thus extending the possibilities of driving a vibrator. Other relevant features are the capability to compensate the perceived intensity of vibration between different actuators or users, and the creation of a C++ API to use the created haptic patterns by an external application.

4. Vibrotactile Stimulation by ERM

Many projects can be found in literature in which vibro-tactile technology is used to transmit haptic information in different areas. This section collects some of research in which ERM type actuators have been used.

4.1 Correction of Posture

One of the advantages of the sense of touch with respect to other senses, such as sight or hearing, is the large number of receptors that are distributed throughout the surface of the human body. This feature allows the brain to associate a specific tactile sensation with a specific spatial point. Many researchers have exploited this fact to develop applications of learning and correction of the posture. Rotella et al. (2012), for example, used five elastic bands with four vibro-tactile motors, each to guide the user to adopt a static posture. The correction of the posture with vibro-tactile information has also been applied to sports (Spelmezan et al., 2009), to learn to play the violin (Van Der Linden et al., 2010), to help in rehabilitation movements (Kapur et al., 2010), or simply to orient the forearm in general learning tasks (Sergi et al., 2008).

4.2 Sensorial Substitution

Vibro-tactile displays can be used to replace or complement the sight or hearing, which is especially important for people with visual or hearing impairments.

Thus, it is possible to provide vibro-tactile environmental information to a blind person in order to avoid obstacles and indicate the path to follow. Zelek et al. (2003) designed a glove with 14 vibrators that represented the information of depth collected with a stereo system of cameras. One of the main keys is the algorithm by which the visual information is matched to the tactile information, since the latter is more limited, not only for the characteristics of the touch, but also for the device used.

4.3 Alerts

Like hearing, and unlike sight, touch is a sense that is always alert to new stimuli. In addition, these stimuli can be of a private nature, which is of great interest for applications, such as the announcement of incoming calls in silent environments by mobile devices, which constitute the most widespread use of vibro-tactile technology. Normally the vibration in the mobile alerts about an event, such as a call or a new message; however, there are applications, such as Vybe, or Contact Vibrate, through which different patterns can be established to provide more information and thus be able to know the sender, for example. In addition, technologies, such as Immersion's VibeTonz are able to accompany the melodies with vibrations, enriching the user experience.

The automotive industry is also benefiting from vibration technology to create warning systems for drivers. Cadillac, for example, has a system of vibro-tactile actuators in the seat that warn the driver of possible road hazards, which are collected through a system of radar, ultrasonic, and vision sensors. Citroën also has a lane-change warning system available, by which it detects when the vehicle is treading a boundary line and vibrates the corresponding seat side.

In scientific literature, there are works, such as Ho et al. (2005), who studied the use of vibro-tactile alerts in the automobile by means of two experiments, concluding that the participants responded more quickly to this type of stimuli than to the visual or auditory ones.

4.4 Navigation and Space Perception

Lindeman et al. (2005) conducted a study to see the effectiveness of applying directional vibro-tactile clues to improve the situational awareness of soldiers in a hypothetical eviction exercise of a building simulated by virtual reality. In this way, the limitations of current technology are also compensated. In this case, for example, directional alerts, in the form of vibrations, help to alleviate the reduced field of vision of a typical virtual reality helmet, informing users about their exposure to areas not cleared and which escape their vision.

The US Aerospace and Naval Medical Research Laboratory created the Tactile Situation Awareness System (TSAS) (Rupert, 2000). This device, which is integrated into the clothing of an aviation pilot, was designed to provide awareness of the spatial situation to pilots using tactors. Thus, the idea is that pilots are able to properly judge the gravity vector or even other flight parameters, such as altitude, speed, or threat situation. One of the versions tested in flight incorporated four columns of five ERM actuators embedded in nylon capsules and placed around the torso. Cardin et al. (2006) did a similar job, focused on informing the pilot when the plane went off course or lost the necessary orientation.

The torso region seems especially suitable for displaying spatial and orientation information by means of a belt (Tsukada and Yasumura, 2004; Ferscha et al., 2008; Nagel et al., 2005). Bloomfield and Badler (2007) used the forearm region to alert the user to collisions in virtual environments.

The seats can also be used to provide spatial information to their user, as proposed by Morrell and Wasilewski (2010) through a 3×5 actuators matrix.

4.5 Sense of Presence

The sensation of presence or immersion is something that has also been tried to improve in the cinema by some researchers, who have explored the possibility of adding a tactile feedback channel. For example, Kim et al. (2009) proposed a data glove that provides viewers with tactile sensations synchronized with audiovisual content. This glove is

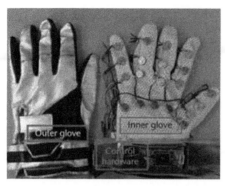 

(a) Control circuit and vibrators of Kim et al. (2009)   (b) Jacket by Lemmens et al. (2009)

Fig. 12: Vibro-tactile devices oriented to increase the sensation of presence.

composed of 20 vibrating motors, whose distribution can be seen in Fig. 12a. According to the authors, when watching a movie, this type of sensations can lead, in addition to greater immersion, to encourage empathy for the characters, helping them to put themselves in their place. These are represented by gray-scale images, such that the gray levels correspond to the intensity of vibration, and their resolution corresponds to the number of actuators.

On the other hand, Lemmens et al. (2009) tried to achieve emotional immersion through a jacket composed of 64 tactile stimulators, as shown in Fig. 12b. The system can thus generate configurable vibration patterns through software and synchronize them with different moments of a movie. According to his hypothesis, tactile stimulation can help to increase emotional immersion, since emotions are often accompanied by different reactions in the body.

4.6 Video Games

Currently the vibro-tactile feedback is present in most commercial video game controls, such as the Nintendo Wii, Microsoft Xbox, and Sony PlayStation in its different versions, or the PC game controls, such as joysticks and gamepads.

Many games take advantage of this technology to create richer experiences, such as driving games, which are capable of recreating complex sensations not only tactile, but of force feedback at the wheel. Companies like Immersion bet heavily on vibro-tactile feedback, creating libraries like their haptic SDK with dozens of effects to be used in games for mobile devices. Some authors go further and propose games that only include tactile content, such as Nordvall (2012) and its Vibrotactile Pong. Orozco et al. (2012) provide more detailed information.

4.7 Telemanipulation and Virtual Reality

Remote manipulation can be defined as the ability of a person to feel and manipulate objects remotely, while telepresence is the ability to make the operator feel at the remote site realistically. In the 70s and 80s, efforts to transmit haptic sensations focused mainly on these latter systems. From the nineties, the haptic term was introduced, and it began to be related to digital environments. Virtual reality systems are nothing but telepresence systems where the remote environment is a digital simulation, so it can benefit from the haptic technology developed for remote manipulation.

Among the devices created for remote manipulation, there are arrays of capacitive pins (Kontarinis et al., 1995), piezoelectric (Debus et al., 2002), moving coils (Murray et al., 2003), (Kontarinis and Howe, 1995; Dennerlein et al., 1997) and ERM actuators (Galambos, 2012).

Cheng et al. (1996) evaluated the use of ERM actuators to replace the return of forces in delicate manipulation operations in virtual environments. The evaluated task consisted of taking a grain of grape and placing it in a glass, and according to its results by tactile return, the time used is reduced compared to sound or visual feedback. It increases instead the pressure exerted, stating that it could be due to overconfidence of the subjects. The device was composed of a manipulator for two fingers composed of two vibrators for each of them, whose intensity is proportional to the force exerted by the user, also informing in turn the collisions that may occur.

Commercially it is possible to find data gloves for virtual reality with vibro-tactile actuators, such as CyberTouch, from CyberGlove Systems (http://www.cyberglovesystems.com). However, they are expensive and with a small number of actuators. That is why many laboratories decide to create their own devices adapted to their needs, such as gloves (Galambos, 2012; Giannopoulos et al., 2012; Sziebig et al., 2009; Graham et al., 2001; Cheng, 1997; Muramatsu et al., 2012; Pabon et al., 2007; Uchiyama et al., 2008), or to be worn on the forearm (Schätzle et al., 2006).

4.8 Other Application Areas

In addition, ERM technology as tactile feedback has been used in other very diverse areas, such as the multimedia enrichment of video scenes (Kim et al., 2009), or the help of musicians, providing information, such as the beat while playing (Hayes, 2011). There are other works related to these actuators. Israr and Poupyrev (2011) proposed an algorithm that makes the user feel the continuous traces in a discrete array of actuators through psychophysical illusions. Cohen et al. (2005) used a piezoelectric film to measure the vibrations of the actuators and thus be able to control the intensity of vibrations by means of a closed-loop control system (Fig. 13). In the experiment, they did not obtain the expected results because the changes in the external conditions that make the vibrators have a smaller amplitude of movement (like the pressure) are compensated with the fact that those same conditions make the user notice more directly actuators, counteracting the effect and noticing a similar subjective intensity. Finally, Van Erp (2002) included a series of recommendations when developing vibro-tactile devices for human-computer interaction.

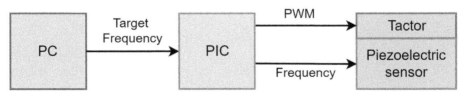

Fig. 13: Closed control loop proposed by Cohen et al. (2005).

5. Conclusions

Touch is still a great challenge in virtual reality research. The tactile and force information people daily get from their hands in the real world is still hard to provide with one single technology. Solutions are often very limited, bulky, or too expensive. When designing the first touch glove at the LoUISE lab several years ago, the team looked for something lightweight and comfortable, but at the same time versatile and affordable. This chapter summarizes that research effort, with an extensive review of the state-of-art in the field of touch in virtual reality, including piezoelectric actuators, alloys with thermal memory effect (SMA), mobile coils, linear resonant actuators (LRA), and vibrating motors (ERM). The main conclusion is that none of the technologies analyzed meets all the requirements to be an ideal touch display, in terms of cost, volume, complexity or resolution. In addition, most of them have other shortcomings, such as their great weight, large volume or structural rigidity that makes them unsuitable to be wearable and use them, for example, in data gloves. Finally, only some of these technologies are commercially available.

After several years of research and prototypes, the team found that vibro-tactile technology was the best that suited these requirements. The result of this long-term work is the current VITAKI (Martínez et al., 2014) platform (Fig. 14, left). This platform has been evaluated in different contexts and tasks. The majority of the evaluations relied on the use of several data gloves. Each of them had a different arrangement of vibro-tactile actuators that were able to provide the specific sensations that were sought, such as the texture (Martínez et al., 2013), weight and size (Martínez et al., 2014, October) or object recognition (Martínez et al., 2014). In addition, this platform has been used in more specific domains, such as physical and cognitive rehabilitation. Thus, VITAKI has been used in the design of new therapies for the treatment of hemi-spatial neglect (Teruel et al., 2015) by offering multimodal stimuli for improving the recognition of the presence of specific object in the 3D space. Moreover, in the rehabilitation domain, it has also been used to provide spatial guidance to visually-impaired people in the execution of cognitive rehabilitation therapies (Oliver et al., 2017; Navarro et al., 2018) (Fig. 14, right). Finally, this platform also analyzed the use of vibro-tactile stimuli in multi-finger exploration of new haptic surfaces (Catalá et al., 2016).

At the LoUISE lab, the team will keep on pushing the possibilities of this technology in the future, and will continue this research work in pursuance of the ultimate solution that makes possible touch in virtual worlds as people perceive it in reality.

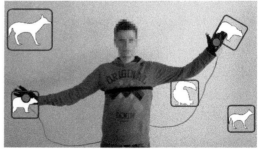

Fig. 14: VITAKI control unit and actuator (left); VITAKI touch gloves used in a rehabilitation exercise (right).

References

Bar-Cohen, Y., Mavroidis, C., Bouzit, M., Dolgin, B.P., Harm, D.L., Kopchok, G.E. and White, R.A. (2001, July). Virtual reality robotic tele-surgery simulations using MEMICA haptic system. pp. 357–364. *In*: Smart Structures and Materials, 2001; Electroactive Polymer Actuators and Devices, vol. 4329, International Society for Optics and Photonics.

Bau, O., Petrevski, U. and Mackay, W. (2009, April). BubbleWrap: a textile-based electromagnetic haptic display. pp. 3607–3612. *In*: CHI'09 Extended Abstracts on Human Factors in Computing Systems, ACM.

Benali-Khoudja, M., Hafez, M. and Kheddar, A. (2007). VITAL: An electromagnetic integrated tactile display. Displays, 28(3): 133–144.

Bliss, J.C. (1969). A relatively high-resolution reading aid for the blind. IEEE Transactions on Man-Machine Systems, 10(1): 1–9.

Bloomfield, A. and Badler, N.I. (2007, March). Collision awareness using vibro-tactile arrays. pp. 163–170. *In*: 2007 IEEE Virtual Reality Conference, IEEE.

Bovenzi, M. (1998). Exposure-response relationship in the hand-arm vibration syndrome: An overview of current epidemiology research. International Archives of Occupational and Environmental Health, 71(8): 509–519.

Brown, L.M. and Kaaresoja, T. (2006, April). Feel who's talking: Using tactons for mobile phone alerts. pp. 604–609. *In*: CHI'06 Extended Abstracts on Human Factors in Computing Systems ACM.

Burdea, G.C. and Coiffet, P. (2003). Virtual Reality Technology, John Wiley & Sons.

Cardin, S., Vexo, F. and Thalmann, D. (2006). Vibro-tactile interface for enhancing piloting abilities during long-term flight (No. Article, pp. p-381).

Carter, T., Seah, S.A., Long, B., Drinkwater, B. and Subramanian, S. (2013, October). Ultra Haptics: Multi-point mid-air haptic feedback for touch surfaces. pp. 505–514. *In*: Proceedings of the 26th Annual ACM Symposium on User Interface Software and Technology, ACM.

Catalá, A., Oliver, M., Molina, J.P. and Gonzalez, P. (2016). Involving multiple fingers in exploring a haptic surface: An evaluation study. The Visual Computer, 32(6-8): 921–932.

Chang, A., O'Modhrain, S., Jacob, R., Gunther, E. and Ishii, H. (2002, June). ComTouch: Design of a vibrotactile communication device. pp. 312–320. *In*: Proceedings of the 4th Conference on Designing Interactive Systems: Processes, Practices, Methods, and Techniques, ACM.

Chen, E.Y. and Marcus, B.A. (1994, November). Exos slip display research and development. pp. 55–1. *In*: Proceedings of the International Mechanical Engineering Congress and Exposition.

Cheng, L.T., Kazman, R. and Robinson, J. (1996, November). Vibro-tactile feedback in delicate virtual reality operations. pp. 243–251. *In*: ACM Multimedia.

Cheng, L.T. (1997, May). Design of a vibro-tactile feedback virtual testbed. pp. 173–176. *In*: CCECE'97, Canadian Conference on Electrical and Computer Engineering, Engineering Innovation: Voyage of Discovery, Conference Proceedings, vol. 1, IEEE.

Cohen, J., Niwa, M., Lindeman, R.W., Noma, H., Yanagida, Y. and Hosaka, K. (2005, April). A closed-loop tactor frequency control system for vibro-tactile feedback. pp. 1296–1299. *In*: CHI'05 Extended Abstracts on Human Factors in Computing Systems, ACM.

Debus, T., Becker, T., Dupont, P., Jang, T.J. and Howe, R.D. (2002, February). Multi-channel vibro-tactile display for sensory substitution during teleoperation. pp. 42–50. *In*: Telemanipulator and Telepresence Technologies, VIII, vol. 4570, International Society for Optics and Photonics.

Dennerlein, J.T., Millman, P.A. and Howe, R.D. (1997, November). Vibro-tactile feedback for industrial telemanipulators. pp. 189–195. *In*: Sixth Annual Symposium on Haptic Interfaces for Virtual Environment and Teleoperator Systems, ASME International Mechanical Engineering Congress and Exposition, vol. 61.

Edmison, J., Jones, M., Nakad, Z. and Martin, T. (2002). Using piezoelectric materials for wearable electronic textiles. pp. 41–48. *In*: Proceedings of Sixth International Symposium on Wearable Computers, IEEE.

Enikov, E.T., Lazarov, K.V. and Gonzales, G.R. (2002, July). Microelectrical mechanical systems actuator array for tactile communication. pp. 551–558. *In*: International Conference on Computers for Handicapped Persons, Springer, Berlin, Heidelberg.

Ferscha, A., Emsenhuber, B., Riener, A., Holzmann, C., Hechinger, M. and Hochreiter, D. (2008). Vibro-tactile Space Awareness.

Frati, V. and Prattichizzo, D. (2011, June). Using kinect for hand-tracking and rendering in wearable haptics. pp. 317–321. *In*: 2011 IEEE World Haptics Conference, IEEE.

Fukuda, T., Morita, H., Arai, F., Ishihara, H. and Matsuura, H. (1997). Micro resonator using electromagnetic actuator for tactile display. pp. 143–148. *In*: 1997 International Symposium on Micromechanics and Human Science (Cat. No. 97TH8311), IEEE.

Galambos, P. (2012). Vibro-tactile feedback for haptics and telemanipulation: Survey, concept and experiment. Acta Polytechnica Hungarica, 9(1): 41–65.

Giannopoulos, E., PomésFreixa, A. and Slater, M. (2012). Touching the void: Exploring virtual objects through a vibro-tactile glove. The International Journal of Virtual Reality, 11(2): 19–24.

Graham, S.J., Staines, W.R., Nelson, A., Plewes, D.B. and McIlroy, W.E. (2001). New devices to deliver somatosensory stimuli during functional MRI. Magnetic Resonance in Medicine: An Official Journal of the International Society for Magnetic Resonance in Medicine, 46(3): 436–442.

Griffin, M.J. (1996). Vibration Discomfort, Handbook of Human Vibration, 988.

Gurari, N., Smith, K., Madhav, M. and Okamura, A.M. (2009, June). Environment discrimination with vibration feedback to the foot, arm, and fingertip. pp. 343–348. *In*: 2009 IEEE International Conference on Rehabilitation Robotics, IEEE.

Hasser, C.J. (1993). Preliminary evaluation of a shape-memory alloy tactile feedback display. pp. 73–80. *In*: ASME, vol. 49.

Hayes, L. (2011). Vibro-tactile feedback-assisted performance. pp. 72–75. *In*: NIME.

Hein, A. and Brell, M. (2007, March). Contact-a vibro-tactile display for computer-aided surgery. pp. 531–536. *In*: Second Joint Euro-Haptics Conference and Symposium on Haptic Interfaces for Virtual Environment and Teleoperator Systems (WHC'07), IEEE.

Ho, C., Tan, H.Z. and Spence, C. (2005). Using spatial vibro-tactile cues to direct visual attention in driving scenes. Transportation Research Part F: Traffic Psychology and Behavior, 8(6): 397–412.

Inaba, G. and Fujita, K. (2006, July). A pseudo-force-feedback device by fingertip tightening for multi-finger object manipulation. pp. 475–478. *In*: Proc. of EuroHaptics.

Ino, S., Shimizu, S., Odagawa, T., Sato, M., Takahashi, M., Izumi, T. and Ifukube, T. (1993, November). A tactile display for presenting quality of materials by changing the temperature of skin surface. pp. 220–224. *In*: Proceedings of 1993 2nd IEEE International Workshop on Robot and Human Communication, IEEE.

Israr, A. and Poupyrev, I. (2011, May). Tactile brush: Drawing on skin with a tactile grid display. pp. 2019–2028. *In*: Proceedings of the SIGCHI Conference on Human Factors in Computing Systems, ACM.

Johnson, D.A. (1990). Shape-memory Alloy Tactical Feedback Actuator, Tini Allow Company, Inc, 2.

Johnson, K. (2002). Neural basis of haptic perception. Stevens' Handbook of Experimental Psychology.

Jungmann, M. and Schlaak, H.F. (2002, July). Miniaturised electrostatic tactile display with high structural compliance. *In*: Proceedings of the Conference on Eurohaptics.

Kajimoto, H., Kawakami, N., Maeda, T. and Tachi, S. (2001, July). Electro-tactile display with force feedback. pp. 95–99. *In*: Proc. World Multiconference on Systemics, Cybernetics and Informatics (SCI2001), vol. 11.

Kajimoto, H., Kawakami, N., Tachi, S. and Inami, M. (2004). Smarttouch: Electric skin to touch the untouchable. IEEE Computer Graphics and Applications, 24(1): 36–43.

Kapur, P., Jensen, M., Buxbaum, L.J., Jax, S.A. and Kuchenbecker, K.J. (2010, March). Spatially distributed tactile feedback for kinesthetic motion guidance. pp. 519–526. *In*: 2010 IEEE Haptics Symposium, IEEE.

Kenaley, G.L. and Cutkosky, M.R. (1989, May). Electrorheological fluid-based robotic fingers with tactile sensing. pp. 132–136. *In*: Proceedings, 1989 International Conference on Robotics and Automation, IEEE.

Kim, K.H., Nam, Y.J., Yamane, R. and Park, M.K. (2009). Smart mouse: 5-DOF haptic hand master using magneto-rheological fluid actuators. *In*: Journal of Physics: Conference Series, 149(1): 012062, IOP Publishing.

Kim, Y., Cha, J., Oakley, I. and Ryu, J. (2009). Exploring Tactile Movies: An Initial Tactile glove Design and Concept Evaluation, IEEE, Multimedia.

King, C.H., Culjat, M.O., Franco, M.L., Bisley, J.W., Dutson, E. and Grundfest, W.S. (2008). Optimization of a pneumatic balloon tactile display for robot-assisted surgery based on human perception. IEEE Transactions on Biomedical Engineering, 55(11): 2593–2600.

Klein, D., Freimuth, H., Monkman, G.J., Egersdörfer, S., Meier, A., Böse, H. and Bruhns, O.T. (2005). Electrorheological tactel elements. Mechatronics, 15(7): 883–897.

Kontarinis, D.A. and Howe, R.D. (1995). Tactile display of vibratory information in teleoperation and virtual environments. Presence: Teleoperators & Virtual Environments, 4(4): 387–402.

Kontarinis, D.A., Son, J.S., Peine, W. and Howe, R.D. (1995, May). A tactile shape-sensing and display system for teleoperated manipulation. pp. 641–646. *In*: Proceedings of 1995 IEEE International Conference on Robotics and Automation, vol. 1, IEEE.

Koo, I.M., Jung, K., Koo, J.C., Nam, J.D., Lee, Y.K. and Choi, H.R. (2008). Development of soft-actuator-based wearable tactile display. IEEE Transactions on Robotics, 24(3): 549–558.

Kyung, K.U., Son, S.W., Kwon, D.S. and Kim, M.S. (2004, April). Design of an integrated tactile display system. pp. 776–781. *In*: IEEE International Conference on Robotics and Automation, 2004, Proceedings, ICRA'04, 2004, vol. 1, IEEE.

Kyung, K.U., Ahn, M., Kwon, D.S. and Srinivasan, M.A. (2005, March). A compact broadband tactile display and its effectiveness in the display of tactile form. pp. 600–601. *In*: First Joint Eurohaptics Conference and Symposium on Haptic Interfaces for Virtual Environment and Teleoperator Systems, World Haptics Conference, IEEE.

Lemmens, P., Crompvoets, F., Brokken, D., Van Den Eerenbeemd, J. and de Vries, G.J. (2009, March). A body-conforming tactile jacket to enrich movie viewing. pp. 7–12. *In*: World Haptics 2009—Third Joint EuroHaptics Conference and Symposium on Haptic Interfaces for Virtual Environment and Teleoperator Systems, IEEE.

Lindeman, R.W., Sibert, J.L., Mendez-Mendez, E., Patil, S. and Phifer, D. (2005, April). Effectiveness of directional vibro-tactile cuing on a building-clearing task. pp. 271–280. *In*: Proceedings of the SIGCHI Conference on Human Factors in Computing Systems, ACM.

Liu, Y., Davidson, R.I., Taylor, P.M., Ngu, J.D. and Zarraga, J.M.C. (2005). Single-cell magnetorheological fluid based tactile display. Displays, 26(1): 29–35.

Mahmoudian, N., Aagaah, M.R., Jazar, G.N. and Mahinfalah, M. (2004, August). Dynamics of a micro electro mechanical system (MEMS). pp. 688–693. *In*: 2004 International Conference on MEMS, NANO and Smart Systems (ICMENS'04), IEEE.

Martínez, J., García, A.S., Molina, J.P., Martínez, D. and González, P. (2013). An empirical evaluation of different haptic feedback for shape and texture recognition. The Visual Computer, 29(2): 111–121.

Martínez, J., García, A.S., Oliver, M., Molina, J.P. and González, P. (2014). VITAKI: A vibro-tactile prototyping toolkit for virtual reality and video games. International Journal of Human-Computer Interaction, 30(11): 855–871.

Martínez, J., García, A.S., Oliver, M., Molina, J.P. and González, P. (2014, October). Weight and size discrimination with vibrotactile feedback. pp. 153–160. *In*: 2014 International Conference on Cyberworlds, IEEE.

Martínez, J., García, A.S., Oliver, M., Molina, J.P. and González, P. (2014). Identifying virtual 3D geometric shapes with a vibrotactile glove. IEEE Computer Graphics and Applications, 36(1): 42–51.

Minamizawa, K., Fukamachi, S., Kajimoto, H., Kawakami, N. and Tachi, S. (2007, August). Gravity grabber: Wearable haptic display to present virtual mass sensation. *In*: ACM SIGGRAPH 2007 Emerging Technologies, p. 8, ACM.

Monkman, G.J. (1992). An electrorheological tactile display. Presence: Teleoperators & Virtual Environments, 1(2): 219–228.

Morrell, J. and Wasilewski, K. (2010, March). Design and evaluation of a vibro-tactile seat to improve spatial awareness while driving. pp. 281–288. *In*: 2010 IEEE Haptics Symposium, IEEE.

Muramatsu, Y., Niitsuma, M. and Thomessen, T. (2012, December). Perception of tactile sensation using vibro-tactile glove interface. pp. 621–626. *In*: 2012 IEEE 3rd International Conference on Cognitive Infocommunications (CogInfoCom), IEEE.

Murray, A.M., Klatzky, R.L. and Khosla, P.K. (2003). Psychophysical characterization and test-bed validation of a wearable vibro-tactile glove for telemanipulation. Presence: Teleoperators & Virtual Environments, 12(2): 156–182.

Nagel, S.K., Carl, C., Kringe, T., Märtin, R. and König, P. (2005). Beyond sensory substitution—learning the sixth sense. Journal of Neural Engineering, 2(4): R13.

Navarro, E., González, P., López-Jaquero, V., Montero, F., Molina Masso, J.P. and Romero-Ayuso, D.M. (2018). Adaptive, multi-sensorial, physiological and social: The next generation of telerehabilitation systems. Frontiers in Neuroinformatics, 12: 43.

Nordvall, M. (2012). SIGHTLENCE: Haptics for Computer Games.

Oliver, M., García, M., Molina, J.P., Martínez, J., Fernández-Caballero, A. and González, P. (2017, June). Smart computer-assisted cognitive rehabilitation for visually impaired people. pp. 121–130. *In*: International Symposium on Ambient Intelligence, Springer, Cham.

Orozco, M., Silva, J., El Saddik, A. and Petriu, E. (2012). The role of haptics in games. *In*: Haptics Rendering and Applications, IntechOpen.

Ottermo, M.V., Stavdahl, O. and Johansen, T.A. (2005, March). Electromechanical design of a miniature tactile shape display for minimally invasive surgery. pp. 561–562. *In*: First Joint Eurohaptics Conference and Symposium on Haptic Interfaces for Virtual Environment and Teleoperator Systems, World Haptics Conference, IEEE.

Pabon, S., Sotgiu, E., Leonardi, R., Brancolini, C., Portillo-Rodriguez, O., Frisoli, A. and Bergamasco, M. (2007, October). A data-glove with vibro-tactile stimulators for virtual social interaction and rehabilitation. pp. 345–348. *In*: 10th Annual Intl. Workshop on Presence.

Pasquero, J. and Hayward, V. (2003, July). STReSS: A practical tactile display system with one millimeter spatial resolution and 700 Hz refresh rate. pp. 94–110. *In*: Proc. Eurohaptics.

Petriu, E.M. and McMath, W.S. (1992). Tactile operator interface for semi-autonomous robotic applications. AIRAS, Artificial Intelligence, Robotics and Automation, Space, pp. 77–82.

Poppinga, B., Pielot, M. and Boll, S. (2009, September). Tacticycle: A tactile display for supporting tourists on a bicycle trip. *In*: Proceedings of the 11th International Conference on Human-Computer Interaction with Mobile Devices and Services, p. 41, ACM.

Rotella, M.F., Guerin, K., He, X. and Okamura, A.M. (2012, March). Hapi bands: A haptic augmented posture interface. pp. 163–170. *In*: 2012 IEEE Haptics Symposium (HAPTICS), IEEE.

Rupert, A.H. (2000). An instrumentation solution for reducing spatial disorientation mishaps. IEEE Engineering in Medicine and Biology Magazine, 19(2): 71–80.

Ryu, J., Jung, J., Park, G. and Choi, S. (2010). Psychophysical model for vibro-tactile rendering in mobile devices. Presence: Teleoperators and Virtual Environments, 19(4): 364–387.

Sarakoglou, I., Tsagarakis, N. and Caldwell, D.G. (2005, March). A portable fingertip tactile feedback array-transmission system reliability and modeling. pp. 547–548. *In*: First Joint Eurohaptics Conference and Symposium on Haptic Interfaces for Virtual Environment and Teleoperator Systems, World Haptics Conference, IEEE.

Sato, M. (2007). Portable pneumatic actuator system using MH alloys, employed as an assistive device. J. Robotics and Mechatronics, 19(6): 612–618.

Schätzle, S., Hulin, T., Preusche, C. and Hirzinger, G. (2006, July). Evaluation of vibro-tactile feedback to the human arm. pp. 557–560. *In*: Proceedings of EuroHaptics.

Scheibe, R., Moehring, M. and Froehlich, B. (2007, March). Tactile feedback at the finger tips for improved direct interaction in immersive environments. *In*: 2007 IEEE Symposium on 3D User Interfaces, IEEE.

Seo, J. and Choi, S. (2010, March). Initial study for creating linearly moving vibro-tactile sensation on mobile device. pp. 67–70. *In*: 2010 IEEE Haptics Symposium, IEEE.

Sergi, F., Accoto, D., Campolo, D. and Guglielmelli, E. (2008, October). Forearm orientation guidance with a vibro-tactile feedback bracelet: On the directionality of tactile motor communication. pp. 433–438. *In*: 2008 2nd IEEE RAS & EMBS International Conference on Biomedical Robotics and Biomechatronics, IEEE.

Shinohara, M., Shimizu, Y. and Mochizuki, A. (1998). Three-dimensional tactile display for the blind. IEEE Transactions on Rehabilitation Engineering, 6(3): 249–256.

Sodhi, R., Poupyrev, I., Glisson, M. and Israr, A. (2013). AIREAL: Interactive tactile experiences in free air. ACM Transactions on Graphics (TOG), 32(4): 134.

Spelmezan, D., Jacobs, M., Hilgers, A. and Borchers, J. (2009, April). Tactile motion instructions for physical activities. pp. 2243–2252. *In*: Proceedings of the SIGCHI Conference on Human Factors in Computing Systems, ACM.

Stevens, S.S. (1959). Tactile vibration: Dynamics of sensory intensity. Journal of Experimental Psychology, 57(4): 210.

Stone, R.J. (2000, August). Haptic feedback: A brief history from telepresence to virtual reality. pp. 1–16. *In*: International Workshop on Haptic Human-Computer Interaction, Springer, Berlin, Heidelberg.

Streque, J., Talbi, A., Pernod, P. and Preobrazhensky, V. (2008, June). Electromagnetic actuation based on MEMS technology for tactile display. pp. 437–446. *In*: International Conference on Human Haptic Sensing and Touch Enabled Computer Applications, Springer, Berlin, Heidelberg.

Summers, I.R., Chanter, C.M., Southall, A.L. and Brady, A.C. (2001, July). Results from a tactile array on the fingertip. *In*: Proceedings of Eurohaptics, vol. 2001.

Sziebig, G., Solvang, B., Kiss, C. and Korondi, P. (2009, May). Vibro-tactile feedback for VR systems. pp. 406–410. *In*: 2009 2nd Conference on Human System Interactions, IEEE.

Talbi, A., Ducloux, O., Tiercelin, N., Deblock, Y., Pernod, P. and Preobrazhensky, V. (2006). Vibro-tactile using micromachined electromagnetic actuators array. *In*: Journal of Physics: Conference Series, 34(1): 637, IOP Publishing.

Tan, H.Z. and Pentland, A. (1997). Tactual displays for wearable computing. Personal Technologies, 1(4): 225–230.

Tang, H. and Beebe, D.J. (1998). A micro-fabricated electrostatic haptic display for persons with visual impairments. IEEE Transactions on Rehabilitation Engineering, 6(3): 241–248.

Taylor, P.M., Hosseini-Sianaki, A., Varley, C.J. and Pollet, D.M. (1997). Advances in an Electrorheological Fluid-based Tactile Array.

Taylor, P.M., Moser, A. and Creed, A. (1997). The Design and Control of a Tactile Display Based on Shape Memory Alloys.

Teruel, M.A., Oliver, M., Montero, F., Navarro, E. and González, P. (2015, June). Multi-sensory treatment of the hemi-spatial neglect by means of virtual reality and haptic techniques. pp. 469–478. *In*: International Work-Conference on the Interplay between Natural and Artificial Computation, Springer, Cham.

Toney, A., Dunne, L., Thomas, B.H. and Ashdown, S.P. (2003). A Shoulder Pad Insert Vibro-tactile Display, p. 35, IEEE.

Tsukada, K. and Yasumura, M. (2004, September). Active belt: Belt-type wearable tactile display for directional navigation. pp. 384–399. *In*: International Conference on Ubiquitous Computing, Springer, Berlin, Heidelberg.

Uchiyama, H., Covington, M.A. and Potter, W.D. (2008, March). Vibro-tactile glove guidance for semi-autonomous wheelchair operations. pp. 336–339. *In*: Proceedings of the 46th Annual Southeast Regional Conference on XX, ACM.

Vallbo, A.B. and Johansson, R.S. (1984). Properties of cutaneous mechanoreceptors in the human hand related to touch sensation. Hum. Neurobiol., 3(1): 3–14.

Van der Linden, J., Schoonderwaldt, E., Bird, J. and Johnson, R. (2010). Musicjacket—Combining motion capture and vibrotactile feedback to teach violin bowing. IEEE Transactions on Instrumentation and Measurement, 60(1): 104–113.

Van Erp, J.B. (2002, July). Guidelines for the use of vibro-tactile displays in human computer interaction. pp. 18–22. *In*: Proceedings of Eurohaptics, vol. 2002.

Verrillo, R.T. (1962). Investigation of some parameters of the cutaneous threshold for vibration. The Journal of the Acoustical Society of America, 34(11): 1768–1773.

Verrillo, R.T., Fraioli, A.J. and Smith, R.L. (1969). Sensation magnitude of vibro-tactile stimuli. Perception & Psychophysics, 6(6): 366–372.

Verrillo, R.T. (1975). Cutaneous sensations. Experimental Sensory Psychology, Glenview: Scott Foresman, 1–10.

Voyles, R., Stavnheim, M. and Yap, B. (1989). Electrorheological Fluid-Based Fingertip with Tactile Sensing, ME319 Class Final Project Report, Stanford University.

Voyles, R.M., Fedder, G. and Khosla, P.K. (1996, April). Design of a modular tactile sensor and actuator based on an electrorheological gel. pp. 13–17. *In*: Proceedings of IEEE International Conference on Robotics and Automation, vol. 1, IEEE.

Wagner, C.R., Lederman, S.J. and Howe, R.D. (2004). Design and performance of a tactile shape display using RC servomotors. Haptics-e, 3(4): 1–6.

Wang, Q. and Hayward, V. (2005, March). Compact, portable, modular, high-performance, distributed tactile transducer device based on lateral skin deformation. pp. 67–72. *In*: 2006 14th Symposium on Haptic Interfaces for Virtual Environment and Teleoperator Systems, IEEE.

Webster III, R.J., Murphy, T.E., Verner, L.N. and Okamura, A.M. (2005). A novel two-dimensional tactile slip display: Design, kinematics and perceptual experiments. ACM Transactions on Applied Perception (TAP), 2(2): 150–165.

Wellman, P.S., Peine, W.J., Favalora, G. and Howe, R.D. (1998). Mechanical design and control of a high-bandwidth shape memory alloy tactile display. pp. 56–66. *In*: Experimental Robotics V, Springer, Berlin, Heidelberg.

Zelek, J.S., Bromley, S., Asmar, D. and Thompson, D. (2003). A haptic glove as a tactile-vision sensory substitution for wayfinding. Journal of Visual Impairment and Blindness, 97(10): 621–632.

Zimmerman, T.G., Lanier, J., Blanchard, C., Bryson, S. and Harvill, Y. (1987, May). A hand gesture interface device. pp. 189–192. *In*: ACM SIGCHI Bulletin, 18(4), ACM.

Index